MOSQUITOES AND DISEASES

MOSQUITOES AND DISEASES

T.V. Sathe
Entomology Division
Department of Zoology
Shivaji University, Kolhapur

&

B.E. Girhe
Department of Zoology
Mudhoji College, Phaltan
Distt. Satara

2018
Daya Publishing House®
A Division of
Astral International Pvt. Ltd.
New Delhi - 110 002

First Published, 2002
Reprinted, 2018

ISBN: 978-81-7035-285-3 (Hardbound)

Publisher's note:

Every possible effort has been made to ensure that the information contained in this book is accurate at the time of going to press, and the publisher and author cannot accept responsibility for any errors or omissions, however caused. No responsibility for loss or damage occasioned to any person acting, or refraining from action, as a result of the material in this publication can be accepted by the editor, the publisher or the author. The Publisher is not associated with any product or vendor mentioned in the book. The contents of this work are intended to further general scientific research, understanding and discussion only. Readers should consult with a specialist where appropriate.

Every effort has been made to trace the owners of copyright material used in this book, if any. The author and the publisher will be grateful for any omission brought to their notice for acknowledgement in the future editions of the book.

Published by : **Daya Publishing House®**
A Division of
Astral International Pvt. Ltd.
– ISO 9001:2008 Certified Company –
4736/23, Ansari Road, Darya Ganj
New Delhi-110 002
Ph. 011-43549197, 23278134
E-mail: info@astralint.com
Website: www.astralint.com

Contents

Preface

Mosquitoes are widespread and diversified taxonomical group of insects and act as important vectors for many diseases like malaria, yellow fever, filariasis, dengue etc. Hence, any advance knowledge on their biodiversity, taxonomy, seasonal abundance and distribution have practical utility in adopting appropriate control measures of mosquitoes and solving the problem related to human health care in the region. In present work, attempts have been made on the biodiversity of mosquitoes with respect to taxonomy, seasonal abundance and distribution. In all, 15 species belonging to the genera *Anopheles*, *Culex* and *Aedes* have been studied. The work presented in this book will be helpful for students, teachers, officers and researchers in Entomology and Epidemiology.

T.V. Sathe

B.E. Girhe

Preface

Chapter 1
Introduction

Inadequate resources of food directly affect the health of a country. The reduction of per capita availability of agricultural land to just 0.1 hector which is totally inadequate to sustain large rural population. Development of agro-based industries and value addition to agro products, however, are essential to increase the per capita income in rural areas, which is just about the fifth to that of urban population, in spite of agriculture contributing to almost 50 per cent of National G.D.P. through direct and indirect means. Starvation of over 200 million people in developing nation was due to poor productivity and due to their incapability to afford the prize of food (Rao, 1996).

The urban population in India, which was barely 30 million in 1900 has already increased to over 240 million, forming almost 30 per cent of our total population and is expected to cross more than 400 million by the year 2001. However, expected resources for sustainable health are not increased upto the satisfaction which leads to affect the health of the nation adversely.

The explosive population growth in urban areas having poor infrastructure and sanitation have turned our large metropolitan cities into mega slums. Over 25 per cent of the urban population in our mega cities still do not have access to safe water supply, making water borne diseases such as cholera, typhoid, tuberculosis, dysentery and gastro-enteritis, the most serious health hazards affecting the urban population. Over 20 per cent of the population in many of our major cities are suffering from hypertension, asthma and other respiratory diseases. Inadequate public transportation and poor sanitation have resulted in brown environmental degradation, converting our major cities into sources of concentrated environmental hazard instead engines of growth. The solid waste generated each day by our mega cities like Calcutta is over 100 tons per million, adding half million ton of solid waste annually, 50 per cent of which is not even collected, let alone recycled, making them the fertile breeding ground for a large number of communicable diseases such as cholera, plaque and malaria.

Education is the most important force for human liberation. It remains the most neglected aspect of our society. Of the global investment of about $1000 billion on education, the total share of three quarters of global population residing in 160 developing nations in the world is a mere, 14 per cent. The total investment on education including technical education and sports from both Central and State Sector in India is less than 3.5 per cent of our G.D.P. or less than $ 25 per capita as against over $ 500 per capita in developed societies (Rao, 1996). In India, still 50 per cent of the school catering can't have a black board, alone library and laboratory facilities. The illiteracy is around 40 per cent of the total population with more than 60 per cent females. The least developed countries with illiteracy rate is about 70 to 80 per cent have only per capita annual income of less than $ 200, while low income nations group like India have a annual income ranging from $ 300 to $ 600 as against over $ 10,000 per capita income of developed nations. The illiteracy rate of India is 35 to 50 per cent where as it is less than 5 per cent in the developed nations. The illiteracy seriously affect the health of a country.

The crude birth rate in Kerala, where female illiteracy is less than 10 per cent has increased to 18 per thousand as against 36 per thousand as States like Uttar Pradesh and Bihar, where female illiteracy exceed 70 per cent. This indicates that there is immediate need to initiate massive illiteracy programme in the States. The reduction in crude birth rate with increase illiteracy in Kerala achieved a low infant mortality rate of less than 17 per thousand as against the prevalent 80 per thousand elsewhere in the country. This clearly indicates that the female literacy, a single factor can avoid the prevalence of diseases in infants. It is therefore, very much essential to have massive education at the grass root level of our society.

The trained man power invariably play an important role in building the health of a country. The number of scientists in India is less than 4.5 per thousand population as against 110 in Japan or over 200 in most of the European nations. The number involved in research which is what really matters in the development of science and technology is just 0.25 per thousand population as against 5 per thousand in all scientifically advanced nations (Rao, 1996). The scientific research can crate necessary technical manpower who can effectively contribute to the wealth of country and technological society of the future in India. The annual expenditure on R and D is less than $ 5 per capita as against over $ 7000 in the developed countries. Large scale tapering with selection process of both teachers and students in our educational system and Govt. employment cannot create excellence.

The large percentage of illiterate females is directly attributable to increased infantile mortality rate as well as poor attention to nutrition and health care. Mortality rate of children, under five, is about 130 per thousand in India as against less than 15 in the U.S. Thought the phenomenal advances achieved in medical sciences and health care, pollution of poverty, widespread illiteracy and environmental degradation in rural areas have resulted in 12 million children in the developing countries dying every year from rampant respiratory and communicative diseases in the world. In India over 3 million children are dying due to above reason. In 8th Plan, India contributed just about Rs. 7500 Crores on health sector which is less than 2 per cent of our plan location.

The poor water supply and sanitation exert pressure on health of a country. In India, eighth planned expenditure on urban water supply, sanitation and other health improvement measures was a meagre 3.5 per cent of the total expenditure of the eighth plan.

Education, environmental management and nutritional security are the basic prerequisites required for promoting health. The development of comprehensive primary health care delivery system is essential to establish the health security in the country. The availability of just about 40 medical practitioners and about 100 beds for a lakh of population is not sufficient. Medical facilities are mostly concentrated in urban areas. Rural medical facilities are totally inadequate as the rural people constitutes 65 per cent of our total population. About 25 thousand primary health care centres and poorly equipped 1.5 lakh of sub centres have been recognized in India, most of which do not have even a skeleton medical staff.

Through space technology, we can today conceive wide spectrum of medical services and assistance to remote areas by providing instructions to rural medics on Pediatrics, internal medicine, orthopedics, ophthalmology, gynecology, nutrition and physical therapy. Timely monitoring and management of environment and eradication of massive illiteracy, particularly female illiteracy would certainly help in solving the problems in health care of diversified society of India. Remarkable developments in medical technology now offer a great promise against several diseases with a genetic basis.

Though recent developments facilitated to cure certain diseases, still there are serious problems in health care. The extraordinary challenges nowadays, are the solution for AIDS and Cancer. The AIDS epidemic is now so wide spread globally that it could help destroy foreign government and contribute to ethnic wars. The World Health Organization says that 23 million people are infected in Sub-Saharan Africa and that new cases are running at about 5,000 a day. The prevalence rate of HIV/AIDS in India is 30.02 per 1000 people surpassed only by Maharashtra and Manipur. The figure has registered an annual 5 per cent increase since 1986 (Thakur, 2000). WHO intimates that the number of Indians with HIV is any where up to 5 million. At present Indian constitute 8 to 10 per cent of global infection with HIV. The incidence of STDs is higher in the coastal areas which subsequently increase the risk of transmission of HIV. About 14 million people acquire STDs every year in India. Effective treatment of this large population requires the use of WHO development syndrome approach to management. Health, education, integration of reproductive and social health and development of an AIDS control programme like "the one is use" in Thailand could help to bring this alarming situation under control remarked by Union Health and F.W. Minister (Thakur, 2000). The other curable diseases are also extraordinary challenge some times in rural and remote areas of India.

Human health is affected by a variety of organisms which include insects, other arthropods, nematodes, protozoans, fungi, bacteria, viruses and rickettsia etc. The pathogens cause several fetal diseases to man likely, insects also cause dreaded diseases to human being. The insects related to epidemiology include houseflies, sandflies, mosquitoes, rat flea and tse-tse fly etc. These insects are vectors of various diseases. Among the above insect vectors, mosquitoes play an important role in epidemiology as they cause fetal diseases like malaria, filariasis, dengue, encephalitis, yellow fever, etc. They also transmit many animal diseases like fowlfox of poultry , mycomatosis of rabbits, Rift valley fever of sheep, encephalitis of horses and birds and heatworms of dogs. Mosquitoes even help transmission of diseases of some other insect, *e.g.* Dermatobia, a South American warble fly transports its eggs to the skin of man and animals through mosquitoes alone. On the skin they hatch and cause myiasis.

Human Malaria is transmitted exclusively by *Anopheles* spp. vectors which include *Anopheles labranchiae* Falleroni, *An. sacharovi* Favre, *An. sargentii* (Theobald), *An. superpictus* Grassi, and *An. pharoensis* Theobald in parts of the mediterranean area, *An. funestus* Giles, *An. moucheti* Evans, *An. nili* (Theobald) and members of the gambiae Giles complex in the Ethiopian region, *An. stephensi* Liston, *An. fluviatilis* Jomes and *An. pulcherrimus* Theobald in Western Asia, *An. culicifacies* Giles in India and Ceylon, *An. maculatus* Theobald, *An. sandaicus* (Rodenwaldf) and members of the *An. barbirostris* Vander Wulp, *An. hyrcunus (pallas)* Wulp groups in Southern Asia; members of the *An. punctulatus* complex in Melayesia and *An. pseudopunctipennis* Theobald, *An. bellator* Dyar and Knab, *An. cnlzii* Dyar and Knab, *An. darlingi* Roof, *An. aquasalis* Curry, *An. albimanus* Wiedemann, *An. albitarsis* Lynch Arribalzaga and *An. nuneztovavi* Gobaldon in Central and South America.

In India, following species of *Anopheles* are most prevalent which transmit malaria. *An. annularis, An. sundaicus, An. stephensi, An. maculatus, An. philippinensis, An. leucosphyrus, An. jluviatilis, An. culicifacies, An. varuna, An. minimus, An. sundaicus.* The prominent species of Culex includes *Culex pipiens fatigans, C. pipiens, C..vishnui, C. pseudovishnui, C. triaeniorhynchus, C. bitaeniorhynchus, C. sinensis, C. gelidus, C. sitiens* and of the Genus *Aedes* are, *Aedes aegypti, A. albopictus, A. vittalis* and *A. variegatus* (Rao, 1984).

Malaria disease is caused by an infection with one or more of several species of blood inhabiting sporozoa of Plasmodium transmitted by the bite of female anopheline mosquitoes. Malaria still is an important disease of man throughout the world and its control is formidable global problem. Malaria is still major killing disease of human of the tropics.

In India, malaria killed 8,00,000 people in 1953 out of 75 million infected people. Thereafter, the cases decline to 60,000 with no death in 1962 but during 1969 the malarial resurgence took place. During

1975, 1976 and 1977 about 5.1,6.4 and 4.7 million people suffered from malaria respectively. However, death due to malaria were 3, 99, 59 and 55 in the year 1974, 1975, 1976 and 1977 respectively. The picture of Maharashtra with respect to incidence of malaria is not encouraging.

Malarial incidence have been studied during the year, 1992-2000 by Girhe and Sathe (2001) with respect to malarial infection cases in human population in Kolhapur district, Maharashtra. It was observed that the incidence of malaria during the years 1992-1996 was in increasing order. Maximum, 700 infection cases have been reported during the year 1996. However, the incidence of malaria declined from the year 1997-2000, showing only 60 infection cases during the year 2000. The reasons for decline in malaria incidence were the proper strategies of mosquito control adopted by the Health Department and low rainfall in the region (Girhe and Sathe, 2001).

Malaria is caused by four species of protozoan namely, *Plasmodium vivax, Plasmodium falciparum* (malignant tertian, subtertian aestivoautumnal malaria, falcifarum malaria); *Plasmodium malariae* (quarten malaria) and *Plasmodium ovale* (ovale malaria). In case of *P. vivax* and *P. falciparum* there are races and strains with respect to clinical picture, geographical distribution and immunological responses. In India, *P. vivax* is responsible for 65-90 per cent where as *P. falciparum* can cause 25-30 per cent of total malaria. The *P. malariae* represent only 1 per cent of the cause where as *P. ovale* is distributed sporadically. The protozoa invade the paranchyma cells of the liver and after passing through developmental stage attach and reside inside R.B.Cs. Female mosquito inject saliva and sporozoites during the act of biting and feeding. Other means of malarial transmission can be either through blood transfusion or congenital transmission or even by the use of a syringe from an infected person.

For completion of life cycle of malarial parasite two different host are required. Asexual reproduction take place in man and sexual in insect vector.

Filariasis is caused by mosquitoes belongs to genera *Culex* and *Mansonia*. Filariasis is a morbid condition caused in man by two nematodes namely, *Wuchereria bancrofti* and *Brngia malayi*. The diseases yellow fever and dengue fever are transmitted by the mosquitoes of Genus *Aedes*, specially Aedes *aegypti* and some other species. Yellow fever is viral fever of short duration varying in severity like toxic jaundice, Albuminuria and haemorrhages etc. According to Atwal (1986) India and South Asia is luckily free from yellow fever virus, however still the possibility of its introduction cannot be ruled out.

The disease dengue is widely distributed in tropical and subtropical areas. The dengue fever is aquite febile illness clinically characterized by haemorrhagic phenomena and tendency to develop a shock syndrome which may be fetal. The Mansonia mosquito belongs to genus Mansonia causes viral encephalitis in man. Japanese encephalitis (J.E.) is an arboviral (B) infection. Recently, in India, Japanese encephalitis incidence was reported from New Delhi, the disease is transmitted by *Culex* species.

Several workers (Mitzmain, 1917; Korke, 1928, 1932; Williams, 1937; Ross *et. al.*, 1943; Puri, 1947, 1948, 1960; Bate, 1949; Sabin, 1952; Horsefall, 1955; Basu, 1958; Soper, 1958; Christophers, 1960; Reid and Knight, 1961, 1966; Mattingly 1962; Reid and Weitz, 1966; Subbarao and Sharma, 1988; etc.) contributed on mosquito borne diseases of man.

The mosquitoes are characterized by small size, 3.00 to 6.00 mm body length (or more), delicate and slender bodied, covered with hairs and scales. They are black or brown, often spotted white and have piercing and sucking type of mouth parts for sucking the blood of animals or cell sap of plants. The larvae of mosquitoes are elongate wrigglers and aquatic in habitat. The pupae are also aquatic and capable of swimming by paddle like movements of the abdomen. The larvae breed in all kinds of fresh water and in the brackish water. The larvae feed mostly on minute algae and other organic matter floating in the water.

Mosquitoes are diversified taxonomical group of insects. Biodiversity of mosquitoes is an important aspect of medical science. Biodiversity was the concept confined to the textbooks and among the biologist for long years. Now, it is figuring as an important item in national and international agenda. We now have a convention on conservation of biodiversity with 159 countries. Biodiversity is variety of living organisms, including their genetic diversity and assemblages they form. In other words, biodiversity entails the interlinked web of life forms ranging from man to microbes. It is the base for continuing evolution which is a phenomenon that has enriched our lives. Biodiversity encompasses the totality of gene species and ecosystems in a region more specially.

Biodiversity is destined to emerge as a new significant and integral aspect of human life. This prodigious diversity is facing a lot of threats and it is a known fact that there are serious losses in terms of both material and knowledge. The global warming, disruption of ozone layer, air, water and soil pollution, acid soils, acid rains, degradation of ground water quality, erosion, pressure on energy, deforestation and depletion of natural resources are the most dangerous ones for concern. Of these, the later two have distinct and direct bearing on the biodiversity (Ramamurty and Ghai, 1993). Man is continuously manipulating 70 per cent of the natural ecosystem for harvesting 90 per cent of his needs (Vitousek *et. al.*, 1986). We have already lost about one million species of plants and animals. The extinction rate is a whopping 150 species per day or 50,000 species per year (Reid and Miller, 1989). At this rate, the loss will be 1 in 100 species per year or 0.7 per cent annually. It is thus feared that more than two million species will be lost in the next decade of which 50 per cent will be arthropods. It would mean that we will lose material along with associated valuable knowledge on ecological and biological issues. However, diversity of fetal mosquito species and diseases are found increasing. Indian species of harmful mosquitoes needs special attention since biodiversity helps in understanding useful and harmful species and different ecosystems and their sustainable use for well being of mankind and nature.

India is considered to be one of the richest centres of biodiversity in the world, particularly because of the large number of diverse ecosystems which nurture an extremely large number of animal and plant species. More than 3 million varieties of plants, animals and micro-organism have been estimated from the world (Rajamani, 1993). Out of the above estimated species, 500,000 have yet been given a scientific name and detailed information is available on about one in hundred only. In India, more than 45,000 species of plants and 75,000 species of animals have been recorded from various parts (Ananthakrishnan, 1993). The insect comprises more than 67,000 species (Nair and Mathew, 1993).

Among the insects, mosquitoes are most important since they are related to health and survival of man. Mosquito biodiversity have been studied by several workers note worthy among them are

Christophers (1933), Barraud (1934), Foote and Cook (1959), William and Maurice (1961), Smith (1973), Busvine (1980), Sathe and Girhe (2001) etc.

There are about 2400 described species of mosquitoes in the world (Stone *et. al.*, 1959; Ward, 1984, 1992). These ubiquitoes insects occur at elevations of 14,000 feet in Kashmir and as low as 3760 feet below sea level in gold mines of South India (Russell *et. al.*, 1943). Larval and adult characters of mosquito are described, keyed and figured (Carpenter and Lacasse, 1955). Biodiversity is not merely the description of different forms and habits. It is study with respect to animal morphology, taxonomy, anatomy, physiology, biochemistry , molecular biology , genetics and genetic engineering, distribution, occurrence etc. In the present study mosquito biodiversity have been studied with respect to taxonomy, distribution and occurrence.

Some workers (Christophers, 1933; Barraud, 1934; James and Liston, 1911; Foote and Cook, 1959; Rao, 1984; Nagpal and Sharma, 1987,1995; Das and Akiyama, 1990; etc.) have contributed on the mosquitoes of India with respect to taxonomy, distribution and seasonal abundance. However, very little attention is paid on mosquitoes of the Maharashtra.

Geographically and climatically Maharashtra (Fig. No.1) is divided into three regions namely, Konkan, Western Ghat and Maharashtra plateau regions. Hence, it would be very interesting to study biodiversity of mosquitoes in these regions and study will be helpful in understanding the epidemiological aspects of the region.

Kolhapur district (Fig. No. 2) is included in Western Maharashtra and is characterized by hilly regions, Western Ghats and several kinds of water bodies. The river Panchganga and Panchganga river system is special attribute of Kolhapur district. It has rainfall ranging from 1500-2700 mm. The above conditions favour mosquitoes for breeding in the area which ultimately leads to infectious diseases like dengue, malaria etc. in human population. Hence, biodiversitical work on mosquitoes of Kolhapur district will add great relevance in health care of the region.

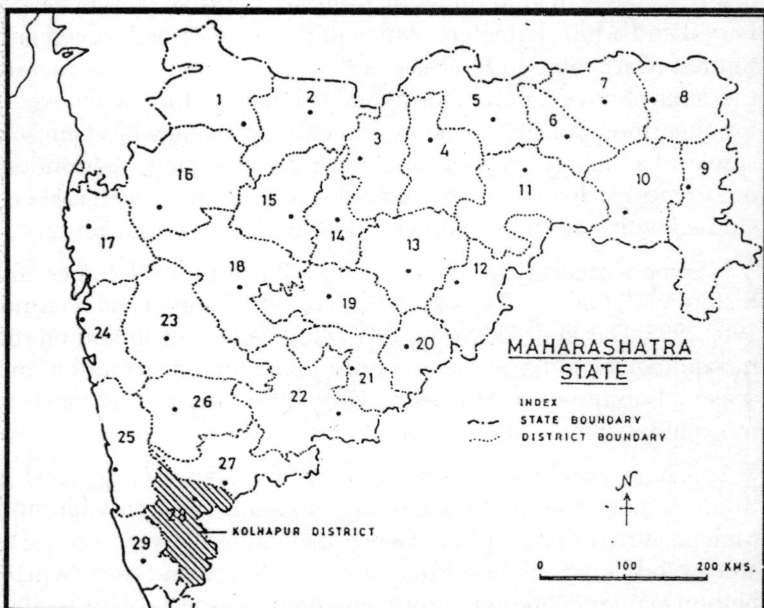

Fig. 1: Map of Maharashtra State

1. Dhule, 2. Jalgaon, 3. Buldhana, 4. Akola, 5. Amaravati, 6. Wardha, 7. Nagpur, 8. Bhandara, 9. Gadchiroli, 10. Chandrapur, 12. Yawatmal, 13. Nanded, 14. Jalna, 15. Aurangabad, 16. Nasik, 17. Thane, 18. Ahmednagar, 19. Bhid, 20. Latur, 21. Osmanabad, 22. Solapur, 23. Pune, 24. Alibag, 25. Ratnagiri, 26. Satara, 27. Sangli, 28. Kolhapur, 29. Kudal.

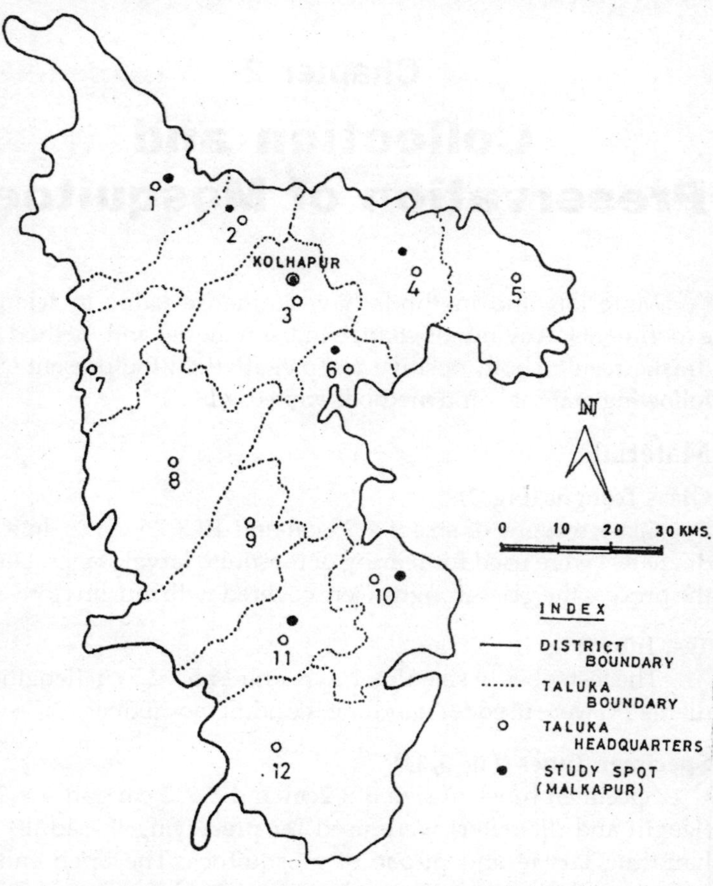

Fig. 2: Map of Kolhapur District Study Area (2, 4, 6, 10, 11)

1. Shahuwadi, 2. Panhala, 3. Karveer, 4. Hatkanangale, 5. Shirol, 6. Kagal, 7. Gaganbavda, 8. Radhanagari, 9. Gargoti, 10. Gadhinglaj, 11. Ajra, 12. Chandgad.

Chapter 2

Collection and Preservation of Mosquitoes

Materials and methods have immense value in scientific experiments. Any minor change in the material and method give drastic results, both positive and negative. In the present study following materials and methods were used.

Materials

Glass Troughs (Fig. 3)

Glass troughs of size 9 × 25 cm and 10 × 25 cm (height and diameter) were used for rearing of mosquito larval stages. During the process the glass troughs were covered with muslin cloth.

Test Tube (Fig. 4)

The test tubes of size 15 × 1.5 cm and 14.5 × 2.5 cm (length and diameter) were used for handling the adult mosquitoes.

Specimen Tubes (Fig. 5, 11)

Specimen tubes of size 6 × 2cm and 5 × 2 cm and 4 × 2 cm (length and diameter) were used for preserving the adults and handling larvae and pupae of mosquitoes. The open ends of specimen tubes were covered either by muslin cloth or wooden cork.

Suction Tube (Fig. 7)

Suction tube has three components; upper glass tube, middle diaphragm and lower rubber tube. The glass tube is 25 cm in length and 1.5 cm in diameter, used for storing the mosquitoes temporarily. Middle diaphragm is a round sieve with 1 mm diameter porosity, the diaphragm is made up of plastic material. Anteriorly, it is attached to glass tube and posteriorly to rubber tube. The diaphragm prevents the entry of mosquitoes towards rubber tube. The third element of this device is a rubber tube of 1.7 cm diameter and 55 cm length. Through this rubber tube, mosquitoes are sucked in glass tube with the help of inhalating air through mouth. The sucked specimen were later separated in other test tubes (Fig. 7).

Insect Net (Fig. 8)

Insect hand net made up of aluminium handle 70 cm long, circular iron ring of 22 cm diameter and ordinary mosquito bag of 70 cm in depth was used for collecting adult mosquitoes.

Material required for collection of larvae and pupae of mosquitoes (Fig. 6a, b, c, d).

Methods

The survey of mosquitoes was made from the district Kolhapur of Maharashtra State (Fig. 1 and 2) from September 2000 to September 2001. A large number of specimen were collected by visiting various places of Kolhapur district namely, Hatkanangale, Kagal, Ajra, Gadhinglaj, Panhala, Jaysingpur and Malkapur at 15 days interval. The mosquitoes were collected with help of suction tube (Fig. 7, 10) and narcotised in ether and killed in cyanide killing bottle. The mosquitoes were pinned with entomological pins from the ventral side and kept on spreading board for drying in the incubator at 60°C. The specimen were pinned in the specimen tube. The dried specimen then kept in sterilized specimen tube by pinning invertedly to wooden cork. A small amount of naphthol was placed at the bottom of the tube and specimen tube was made air tight (Fig. 5). The whole mount of some specimen have also been made on the slide and the slides are kept in slide boxes. The records were made on locality, date of collection and identification. For the details, head, antenna, wing, legs and abdomen were mounted on slide in D.P.X. (Fig. 10). Morphological studies were carried out with the help of monocular microscope. Comparative measurements of body and body parts of specimen were made with ocular micrometer and calculated with the help of graduated mechanical stage. All measurements were made in millimeter.

The collection of mosquitoes were identified, belongs to different families proposed by Christophers (1933), Barraud (1934), Foote and Cook (1959), Rao (1984) and Nagpal and Sharma (1995). A large number of references were consulted in the course of the studies and cited in the text.

The terminology adopted for description of species is same as that of Christophers (1933), Barraud (1934) and Tanaka (1979) which is represented in the figures 12 to 24.

Photography

The whole mounts of the mosquitoes were considered for microscopic photography. The coloured photography of whole mounts was made with the help of Asahi Pentex Auto Bellows K set Camera and Orthopland microscope (Germany).

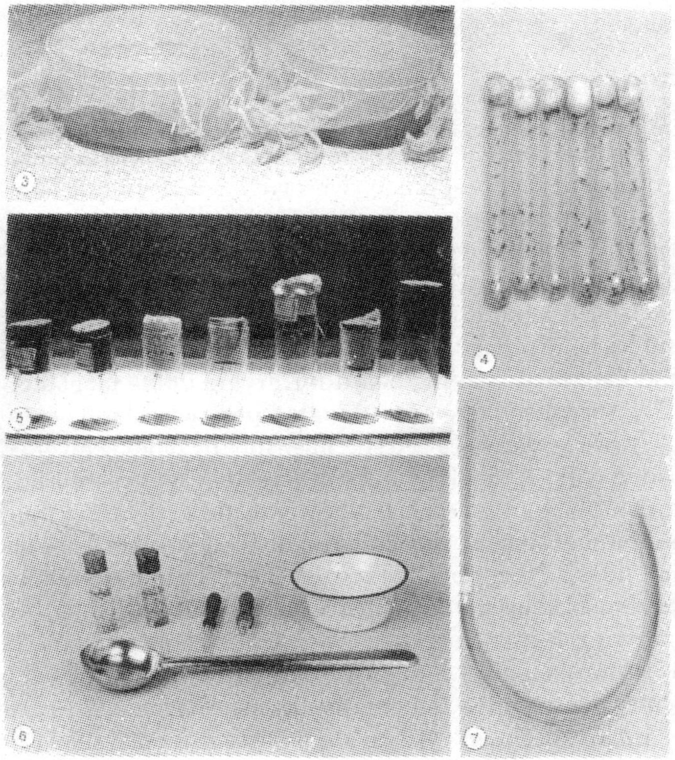

Plate I

Fig. 3: Glass Troughs, Fig. 4: Test Tubes, Fig. 5: Specimen Tubes, Fig. 6: Material Required for Collection of Larvae and Pupae of Mosquitoes, Fig. 7: Suction Tube.

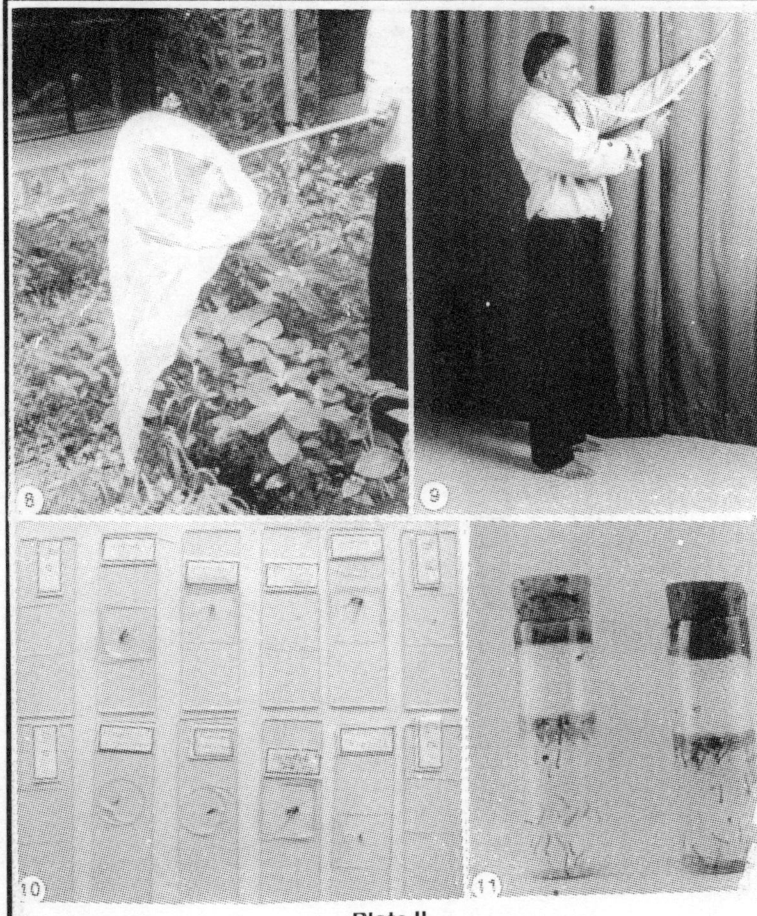

Plate II

Fig. 8: Insect Net, Fig. 9: Collection of Mosquitoes Through Suction Tube, Fig. 10: Whole Mount on Slide, Fig. 11: Larvae and Pupae in Specimen Tubes.

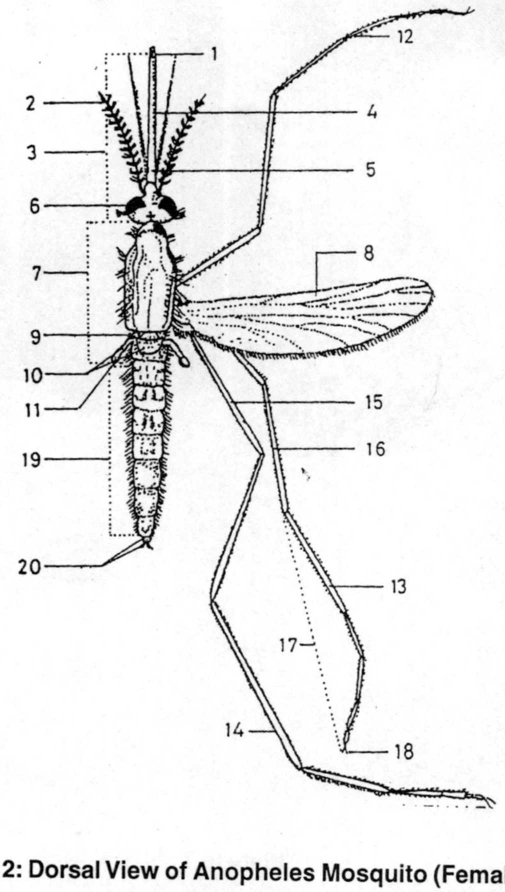

Fig. 12: Dorsal View of Anopheles Mosquito (Female)

1. Labellum, 2. Antenna, 3. Head, 4. Proboscis, 5. Maxillary palp, 6. Eye, 7. Thorax, 8. Wing, 9. Scutellum, 10. Post scutellum, 11. Halter, 12. Foreleg, 13. Midleg, 14. Hindleg, 15. Femuer, 16. Tibia, 17. Tarsus, 18. Claw, 19. Abdomen, 20. Cerci.

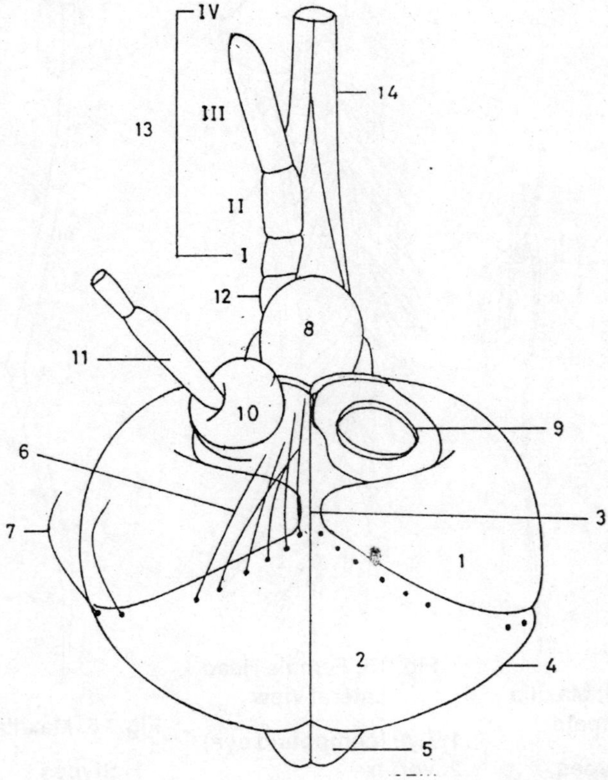

Fig. 13: Female Head Dorsal View

1. Eye (compound eye), 2. Vertex, 3. Interocular space, 4. Tempus, 5. Nape, 6. Verticle bristles, 7. Temporal bristles, 8. Clypeus, 9. Scape (Antenna), 10. Pedicel (Antenna), 11. Flagellomere (Antenna), 12. Palpifer, 13. Palpus, 14. Proboscis.

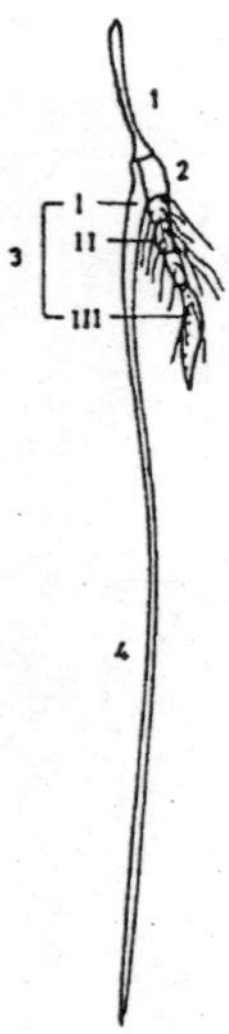

Fig. 14: Maxilla Female

1. Stypes
2. Palpifer
3. Palpus
4. Galea

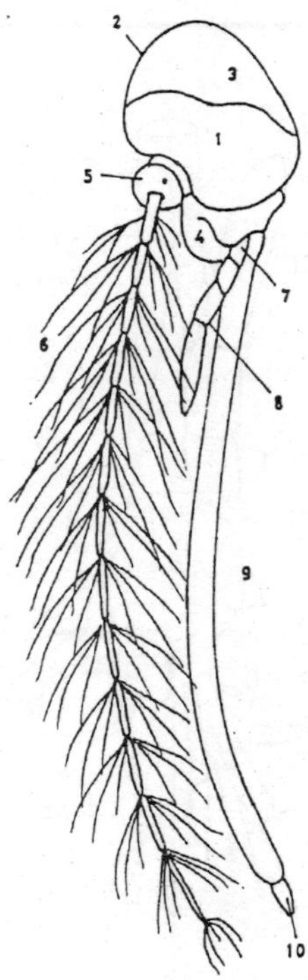

Fig. 15: Female Head Lateral View

1. Eye (compound eye)
2. Vertex
3. Tempus
4. Clypeus
5. Pedicel (Antenna)
6. Flagellum (Antenna)
7. Palpifer
8. Palpus
9. Labium (Proboscis)
10. Labellum (Proboscis)

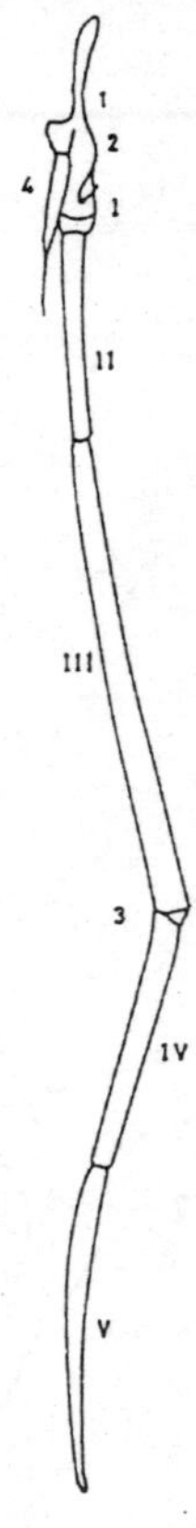

Fig. 16: Maxilla Male

1. Stypes
2. Palpifer
3. Palpus
4. Galea

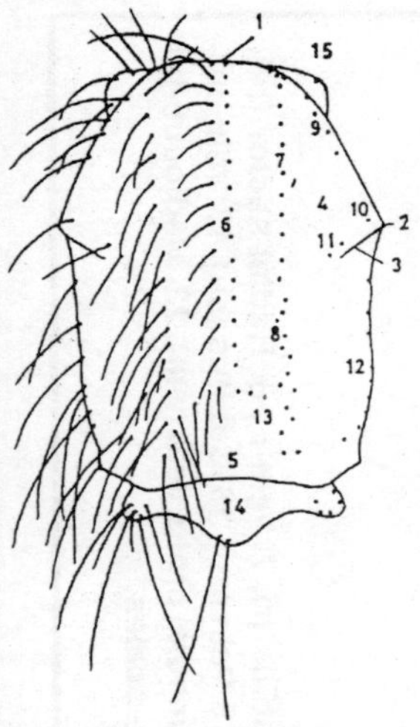

Fig. 17: Thorax Dorsal View

1. Anterior pronotory, 2. Scutal angle, 3. Scutal suture, 4. Fossal area, 5. Prescutellar space, 6. Acrostichal bristles, 7. Anterior dorsocentral bristles, 8. Posterior dorsocentral bristles, 9. Humeral bristles, 10. Angular bristles, 11. Posterior fossal bristles, 12. Subaalar bristles, 13. Prescutellar bristles, 14. Scutellum, 15. Anterior pronotal lobe.

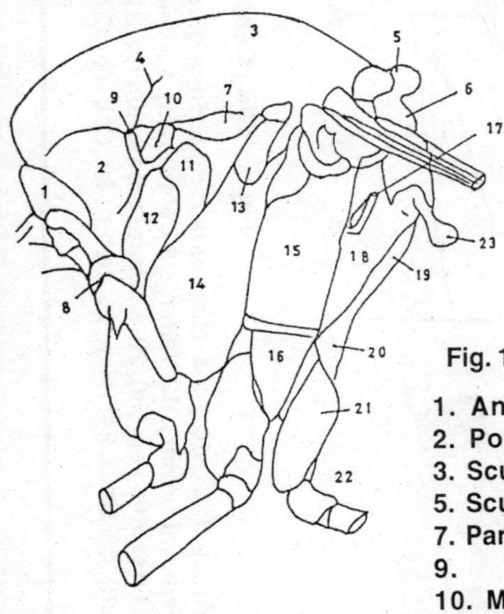

Fig. 18: Thorax Lateral View

1. Anterior pronotal lobe, 2. Posterior pronotal lobe, 3. Scutum, 4. Scutal suture, 5. Scutellum, 6. Post notum, 7. Paratergite, 8. Propleuron, 9. Spiracular area, 10. Mesothoracic spiracle, 11. Post spiracular area, 12. Subspiracular area, 13. Prealar knob, 14. Sternopleuron, 15. Mesepimeron, 16. Mesomeron, 17. Metathoracic spiracle, 18. Metepisternum, 19. Metepimeron, 20. Metameron, 21. Hindcoxa, 22. Hindtrochanter, 23. Halter, 24. Wing.

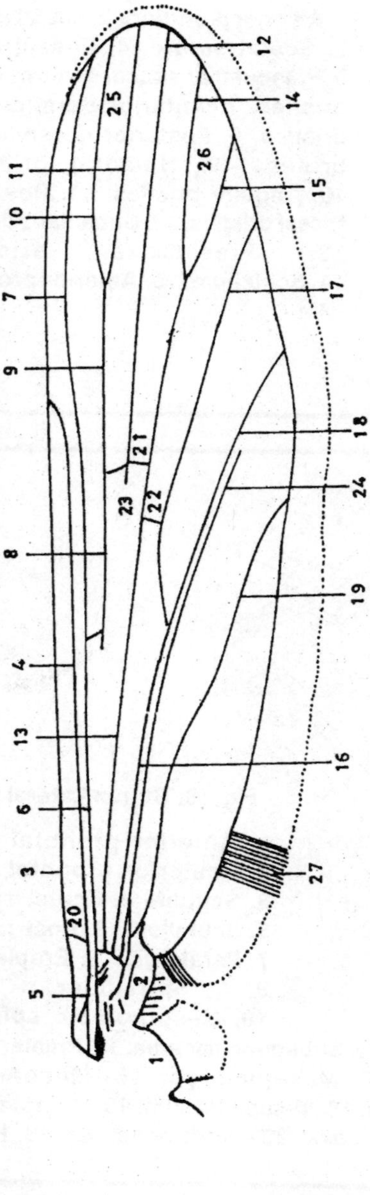

Fig. 19: Wing General

1. Squama, 2. Alula, 3. Costa (c), 4. Subcosta (sc), 5. Remigium, 6. Radius (r), 7. Vein r_1, 8. Radial sector (rs), 9. Vein r_{2+3}, 10. Vein r_2, 11. Vein r_3, 12. Vein r_{4+5}, 13. Media (m), 14. Vein m_{1+2}, 15. Vein m_{3+4}, 16. Cubitus (cu), 17. Vein cu_1, 18. Vein cu_2, 19. First anal vein (la), 20. Humeral cross vein (h), 21. Radiomedial cross vein (r-m), 22. Mediocubital cross vein (m-cu), 23. Spur, 24. Plica, 25. Cell r_2, 26. Cell m_{1+2}, 27. Fringe scales.

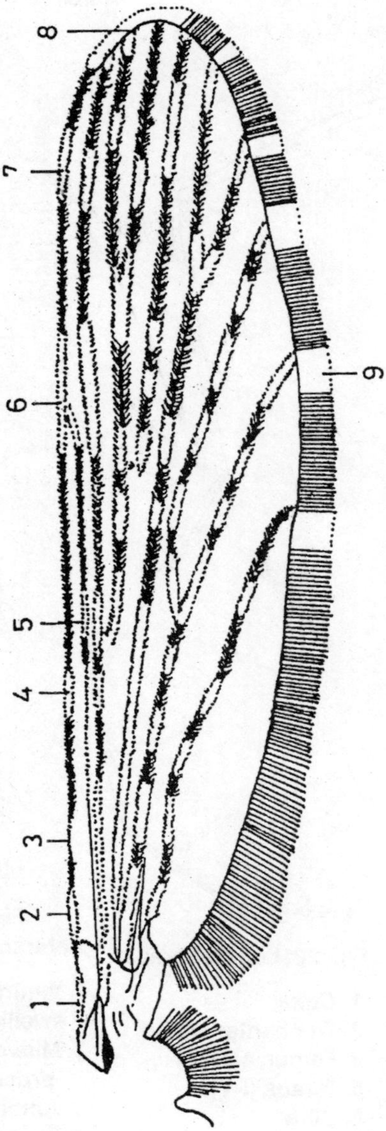

Fig. 20: Wing of Anopheles

1. Prehumeral, 2. Humeral, 3. Presector, 4. Sector, 5. Accessory sector, 6. Subcostal, 7. Preapical, 8. Apical, 9. Fringe.

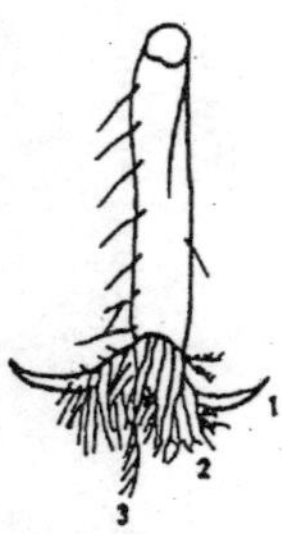

Fig. 21: Foretarsomeres Female

1. Claw
2. Pulvillus
3. Empodium

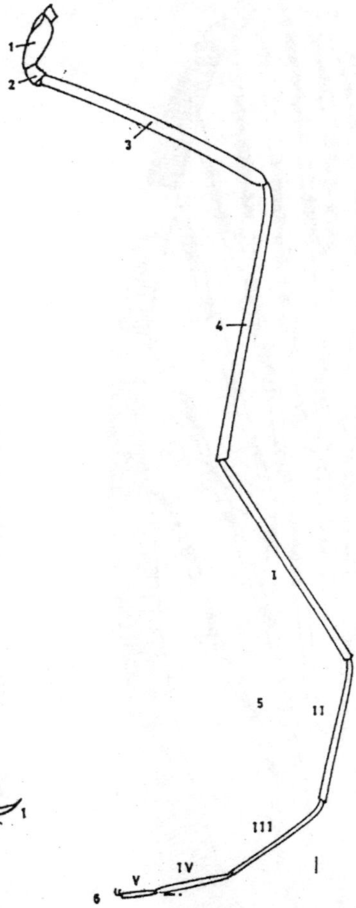

Fig. 22: Leg General

1. Coxa
2. Trochanter
3. Femur, 4. Tibia
5. Tarsus (I-V)
6. Claw

Fig. 23: Foretarsomeres Male

1. Ventrobasal swelling
2. Midventral process
3. Anterior claw
4. Posterior claw
5. Empodium

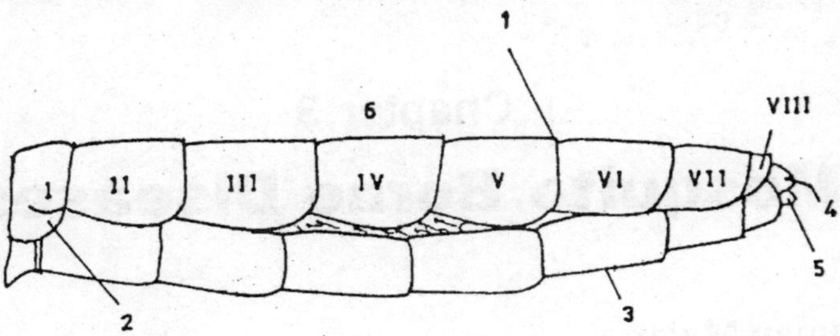

Fig. 24: Abdomen Lateral View

1. Tergum, 2. Laterotergite, 3. Sternum, 4. Cercus, 5. Postgenital plate, 6. Abdominal Segments (I-VIII).

Chapter 3

Mosquito Borne Diseases

Human Malaria

Disease

Malaria (ague, paludism, marsh fever, intermittent fever, wechselfieber, kaltesfieber).

Symptoms

Acute illness with chills, fever, Headache, stiffness of joints etc.

Medical Treatment

Antibiotics: Chloroquin, Arthemether.

Pathogen

Protozoa, sporozoa, plasmodium: *Plasmodium vivax, P. falciparus, P. malariae, P. ovale* and *P. falciparus* have greatest importance.

Vectors

Anopheles mosquitoes.

A. subpictus (India, Indonesia, Java, Andman and Nicobar, Lakshadweep, China, Afganistan etc.)

A. culicifacies (India, Afghanistan, Iran, Myanmar , Nepal, Sri Lanka, Pakistan, Bangladesh, China, Thailand, Oman etc.)

A. stephensi (India, Iran, Iraq, Bangladesh, Afghanistan, Pakistan, Nepal, Myanmar, Thailand, China, Indonesia etc.)

A. sinensis (China, Java, Sumatra)

A. fluviatilis (India, Iran, Pakistan)

A. minimus (India, Nepal, Myanmar etc.)

A. varuna (India)

A. sundaicus (India, China, Indonesia, Myanmar, Thailand, Malaysia etc.)

A. philipinensis (India, Bangladesh, Myanmar etc.)

A. annularis (India {Orissa})

Reservoirs

Man, frequently marginally symptomatic. Nonhuman primate reservoirs are known as monkey or suspected for two human plasmodia but, are not required to maintain human disease.

Distribution

Mainly humid tropics between 45°N and 40°S latitude; especially Africa, Asia, Central and South America. Formerly wide spread in temperate Europe and North America.

More important *Anopheles* vectors of malaria:

North America: (a) Southeastern – *quadrimaculatus,* (b) Southwestern – *freeborni,* (c) Mexico – *albimanus, aztecus, pseudopunctipennis.*

Central America and West Indies: *albimanus, aquasalis, bellator, pseudopunctipennis, punctimacula.*

South America: *albimanus* (Ecuador, Colombia, Venezuela), *albitarsis, aquasalis, bellator, cruzii, darlingi, nuneztovari* (Northern), *pseudopunctipennis* (Northern and Western), *punctimacula.*

North European and Asiatic: *atroparvus, messeae, pattoni* (northerna China), *sacharovi, sinensis* (southern China). Mediterranean – Southern Europe; Morocco, Algeria, Tunisia; Through the Levant to the Sea of Aral : *atroparvus* (Spain, Portugal), , *claviger, dthali, labranchiae, messear, pulcherrimus, sacharovi.*

Desert: North Africa and Arabia: *hispaniola, multicolor, pharoensis, sergentii.*

Ethiopian

(a) African – *dthali, funestus, gambiae* (sp. A, B) + *melas* (West Coast), *merus* (East Coast) , *moucheti, nili, pharoensis.*

(b). Yemen – *culicifacies, gambiae, sergentii.*

Indo-Persian – Iraq, Oman, Persia, Afghanistan, Pakistan, India, Sri Lanka: *annularis, culicifacies, dthali, fluviatilis, hyrcanus, minimus, philippinensis, pulcherrimus, stephensi, sundaicus, superpictus, varuna.*

Indochinese Hill Zone – Foothills of Himalayas to Hills of S. China, Burma, Thailand, and Indo-China: *annularis, balabacensis, maculatus, minimus.*

Malaysian – Malaya, Indonesia, Borneo, Philippines, Coastal Plains from S. China to Bengal: *aconitus, balabacensis, campestris, donaldi, flavirostria* (Philippines), *letifer, leucosphyrus, maculatus, minimus, philipinensis, sinensis, subpictus, sundaicus.*

Chinese – Central China, Korea, Japan: *lesteri, pattoni, sacharovi, sinensis.*

Australasian : *bancrofti, farauti, kanvari, koliensis, punctulatus, subpictus.*

Human Filariasis

Disease

Bancroftian filariasis, brugian filariasis (lymphatic filariasis, obstructive filariasis, elephantiasis).

Symptoms

Cough, Itching etc.

Medical Treatment

Carvamazine

Pathogen

Filarial nematodes

Wuchereria bancrofti and *W. (=Brugia) malayi*

W. malayi is transmitted mainly by *Mansonia* and *Anopheles* mosquitoes and *W. bancrofti* is transmitted by various species of *Anopheles, Culex* and *Aedes* mosquitoes.

Vectors

Several species of night or day active mosquitoes : especially *Culex pipiens, quinquefasciatus, Mansonia annulifera, M. uniforms.* In India, *M. annulifera* is chief vector.

Reservoirs

Bancroftian fillariasis : Man; *Brugian filariasis* : Man and variety of mammals (Sub-periodic strain).

Distribution

Humid tropics and around the Mediterranean.

Importance

At least 250 million cases have been recorded worldwide. Debilitation continuously affects the economics of underdeveloped countries.

Yellow Fever

Disease

Yellow fever (YF) , sylvatic and urban.

Symptoms

Cough, headache, pain in joints.

Medical Treatment

Symptomatic treatment.

Pathogen

A virus belongs to family Togaviridae *Flavivirus*

Victims

Humans, monkeys

Vectors

Urban form, *Aedes aegypti.* Sylvatic form, Africa forest canopy and perydomestic *Aedes.* South and Central America forest canopy, *Haemagogus* and *Sabethes.*

Reservoirs

Partially resistant species of monkeys. There is continuous transmission by mosquitoes among susceptible species of monkeys.

Distribution

Humid tropics of Africa, Central and South America. Urban form, once wide spread, wherever *Aedes aegypti* occurred, except never established in Asia.

Importance

Formerly a cause of epidemics in urban centres of Africa, the New world and Europe. Presently about 200 clinical cases of sylvan YF are reported annually from Africa and South America with about 30 to 75 percent mortality and undoubtedly more cases occur.

Dengue (Den)

Disease

Dengue (DEN 1, 2, 3, 4) dengue hemorrhagic fever, dengue shock syndrome, epidemic hemorrhagic fever, breakbone fever.

Symptoms

Fever, body ache, bleeding in nose.

Medical Treatment

Symptomatic treatment.

Pathogen

A virus belongs to family Togaviridae, *Flavivirus;* four distinct serotypes.

Victims

Humans, hemorrhagic and shock manifestations especially serious in infants and children.

Vectors

Aedes mosquitoes in sub-genus *Stegomyia specially A. aegypti.*

Vector species

Aedes aegypti, A. albopictus.

Reservoirs

Continuous transmission by *Stegomyia* between humans, other mechanisms not substantiated.

Distribution

Coincides with *A. aegypti* within about 40°N and 45°S latitude. Currently mainly in South-east Asia and Caribbean.

Importance

Outbreaks of debilitating but non-fated disease, may involve a half million to two million humans. Large urban centres in Asia, particularly Bankok, experience frequent hemorrhagic outbreaks with notable mortality, especially in children.

Japanese Encephalitis

Disease

Japanese encephalitis (JE, JBE).

Symptoms

Bleeding in nose, fever, body ache.

Medical Treatment

Symptomatic treatment.

Pathogen

A virus (Togaviridae) , *Flavivirus*

Victims

Humans.

Vectors

Culex mosquitoes: especially *C. tritaeniorhynchus* in Japan and certain other *Culex* elsewhere.

Reservoirs

Uncertain, may include birds; pigs are amplifying hosts.

Distribution

From eastern Siberia westward across Asia to India.

Importance

Epidemis irregular, involved upto around 6,000 clinical cases with 3C percent mortality. Permanent impairment known.

Chapter 4
Biodiversity: Taxonomy

Taxonomy was treated unrewardly and stepmotherely by the financial agencies and secondly, the students find field like molecular biology. Hence, it has never been an attractive profession but, now the picture is very fastly changing in favour of taxonomy since taxonomists need in research institutes, museums, central and government agencies, industries, zoos, universities etc. Taxonomy plays a crucial role in the field of applied biology including public health, national defence, pest management, environmental problems, wild life management, nutritional science, forensic science and several other fields in identifying the species. In fact, nowadays, taxonomists are largely involved in designing and implementing control programmes of pest and diseases effectively. Taxonomy and other branches of biological sciences are inter-linked. Taxonomy also helps in evolutionary history of the species and guide the explorations for native and exotic species for fruitful use.

The mosquitoes are of great hygienic importance and economic losses due to mosquitoes no doubt, amply justify the great sums. Hence, in the present topic attempts have been made on taxonomy of the mosquitoes. In past, Strickland and Choudhury (1927), Knowles and White (1927), Christophers (1933), Barraud (1934), Afridi and Puri (1940), Puri (1948, 1960), Wattal and Karla (1961), Roy and Brown (1970), Russell *et. al.* (1943), Stone *et. al.* (1959), etc. have contributed on the taxonomy of the mosquitoes. Christophers (1933), Barraud (1934), Evans (1938), Stone *et. al.* (1959), Ross and Robert (1943), Wattal and Karla (1961) Harrison and Scanlon (1973), Knight and Stone (1977), Rao, (1984) etc. contributed on Indian mosquitoes.

The mosquitoes belong to the family Culicidae of Order Diptera. The family Culicidae is characterized by following features:

1. Vein subcosta (SC) reaching the costa, R_{2+3} forked, R_{4+5} simple, no cross-vein connection of R_1 and R_2, media (M) two branched and anal vein (A) long and reaching wing margin.
2. Scales present on wing, head and body.
3. Absence of a definite suture between the prescutum and scutum, pronotum completely divided.
4. Mouth parts elongated, formed for piercing and for blood sucking in the female, males not blood suckers.
5. Antennae long and filamentous, with 14 and 15 segments, in male brushy and plumose while, in female hairy and pilose.
6. Larvae aquatic, with complete head capsule, one pair of functional spiracles situated dorsally on eighth abdominal segment.

The family Culicidae contain three sub families *viz.*,

1. Toxorhynchitinae
2. Anophelinae
3. Culicinae

The sub family Toxorhynchitinae is represented by following characters:

1. Proboscis long, broad at base and flexible distally.
2. Adult being flower feeding.
3. Larvae predaceous.

As the members of subfamily Toxorhynchitinae do not posses blood sucking proboscis, they can not feed on animals. Therefore, medically this subfamily has no importance.

The subfamily Anophelinae is recognized by following characters:

1. Palpi of male and female as long as proboscis (except *Bironella)*.
2. Scutellum rounded.
3. Leg very long and slender.
4. Tibial bristles and pulvilli absent.

5. Abdomen without scales.
6. Wing usually with light and dark markings.

According to Edwards (1932) and Culicidologists the subfamily Anophelinae is divided into three genera, namely *Chagasia* Cruz, *Bironella* Theobald and *Anopheles* Meigen. The genus *Chagasia* Cruz is recognized by sculellum slightly trilobed; *Bironella* Theobald with scutellum rounded, wing with stem of median (M) forked and wavy and *Anopheles* Meigen with scutellum evently rounded and wing with stem of median (M) fork straight.

The subfamily Culicinae is medically very important it shows following characters:

1. Scutellum trilobed, each lobe bearing bristles and area between the lobe without bristles.
2. Abdomen completely clothed with broad scales which always lie flat.
3. Scales present on veins as well as along hind margin of wing.
4. Mouth parts long and in female adopted for piercing.
5. In male, palpi long and with uniform thickness throughout the entire length, terminal part generally with little bend.
6. In female, palpi conspicuously shorter than the proboscis.

The subfamily Culicinae includes about 1500 species distributed to more than 20 genera. Two third of the described species belongs to the genera *Culex* and *Aedes*.

GENUS ANOPHELES MEIGEN 1818

Meigen. Syst. Besshr. I.p. 10, 1818 (*Anopheles* Hgg.)

The genus *Anopheles* is erected by Meigen in 1818. This genus shows following features :

1. Body not hump backed when at rest.
2. In female, palpi as long as proboscis and slender.
3. In male, palpi club shaped.
4. Palpi five segmented.
5. Proboscis in straight line with body.
6. Pronotum lack setae.

7. Scutellum half moon shaped or rounded with a uniform row of hairs along the margin.
8. Wing spotted with white and dark scales.
9. Abdomen densely hairy.
10. Mandibles and maxillae of female well developed and toothed.
11. Anal cerci very small.

Four hundred and twenty species have been reported under the genus *Anopheles* from the World, out of which fifty are well known vectors of malaria. Christophers (1933) divided the genus *Anopheles* into four subgenera,

1. *Anopheles,*
2. *Myzomyia,*
3. *Nyssorhynchus* and
4. *Stegomyia.*

However, recently, Nagpal and Sharma (1995) visualized six subgenera namely,

1. *Anopheles* Meigen,
2. *Cellia* Theobald,
3. *Kerteszia* Theobald,
4. *Lophopodomyia* Antunnes,
5. *Nyssorhynchus* Blanchard and
6. *Stethomyia* Theobald of this genus.

They visualized twelve zones of epidemiology of the World. According to them the subgenus *Anopheles* found in all the twelve zones with a maximum distribution in Malaysian (56 species) and minimum from Afrotrophical zones (11 species). The subgenus *Cellia* is found in 10 zones and a maximum number of species are recorded from Afrotrophical zone (115 spp.) and a minimum from South American zone (1 sp. only). The subgenus *Kerteszia* is reported from only three American zones, with a maximum number from South America (11 spp.) followed by Central America (5 spp.) and North America (1 sp. only). The subgenus *Lophopodomyia* is reported from two *zones*, South America (6 spp.) and Central American (1 sp. only). The subgenus *Nyssorhynchus* is reported only from four epidemiological zones with a maximum distribution in South America (30 spp.,) followed by Central American (9 spp.), North American (4 spp.) and Chinese zone (1 sp.). The subgenus *Stethomyia* is reported from two regions only, South America (5 spp.) and Central America (2 spp.). From the subgenus *Cellia* 197 species have been reported from the world and 34 species from India (Nagpal and Sharma, 1995).

From India, 58 species of the genus *Anopheles* have been reported (Rao, 1975; Nagpal and Sharma, 1995). From Kolhapur district, Maharashtra four species of *Anopheles* have been reported (Sathe and Girhe, 2001) and have been redescribed in the text.

ANOPHELES (CELL IA) CULICIFACIES GILES

Female (Fig. 25)

5.2 mm long, 1.2 mm broad; antenna 1.45 mm long, pilose; fore wing 3.55 mm long, hind leg 9.85 mm long, head dark brown, thorax dark brown, abdomen reddish brown.

Head

0.5 mm long, 0.6 mm broad, globular; vertex smooth, dark brown; eyes large, rounded and black; interocular space 0.16 mm, clypeus broad, convex, brownish; proboscis 2.1 mm long, cylindrical, straight and black; palpi pale brown, slender as long as proboscis, apical pale band of palpi nearly equal to pre apical dark band, no pale scales on palpi; mandibles and maxillae yellow, well developed, toothed; antenna 15 segmented, 1.45 mm long, hairy, pilose and dark brown; scape 0.20 mm long, 0.17 mm broad; pedicel 0.09 mm long, 0.14 mm broad, brown; flagellum 13 segmented, dark brown, 1.16 mm long.

Flagellar formula

1 L/W = 0.15, 13 L/W = 0.25, L 1/13 = 0.060, W 1/13 = 0.10.

Thorax

1.3 mm long, 1.15 mm broad, laterally compressed, dark brown; scutum shield like; anterior pronotal lobe brown, sparesely hairy; mesothoracic spiracles rounded with bristles, postnotum yellow and rounded, prealar knob and sternopleuron pale yellow, scutellum rounded with a uniform row of hairs along the margin; sternopleuron and mesepimeron flat, nearly rectangular, pale yellow.

Fore wing

3.55 mm long, 0.91 mm broad, spotted with white and dark scales; scales 0.02 mm long, elongated, present on veins as well as fringe on hind margin of wing; squama and alula semicircular, transparent with scales on the hind margin; R_{2+3} forked, R_{4+5} straight, subcosta straight, 2.90 mm long, media straight and bifurcated, 2.60 mm long, radius straight and slightly curved at apex, 3.40 mm long; no cross-vein connection of R_1 and R_2, cross-veins r-m and m-cu both present; cubitus bifurcated, cu_1 1.68 mm long; anal (A) vein 1.51 mm long and reaching margin; wing with four dark spots on costa involving vein 1, vein 3 (R_{4+5}) mainly dark, inner costa interrupted.

Angle between m-cu and cu_1 90°, m_{3+4} and r-m 85° and R_{2+3} and r-m 35°.

Halter

0.25 mm long, 0.21 mm broad, dark brown, rounded at tip.

Hind Leg

9.85 mm long, slender without narrow white bands; coxa 1.20 mm long, broad, yellowish brown with bristles; trochanter 0.35 mm long, hard and dark brown; femur 2.00 mm long, cylindrical, yellow; tibia 2.3 mm long, slender and curved; tarsus 4.00 mm long, 5 segmented and without bands; femur and tibia not speckled, tip of hind leg not white.

Other legs

Special marks: Similar.

Abdomen

3.4 mm long, 1.2 mm broad, reddish brown, cylindrical, unbanded, covered with densely fine hairs; scales absent, abdomen tip pointed; anal cerci 0.15 mm long, 0.11 mm broad, slender very small.

Colour

Black	:	Eye, proboscis.
Brown	:	Palpi, antenna, scape and pedicel.
Yellow	:	Pronotum, maxillae, mandibles.

Male

4.8 mm long, slender, smaller than female, phytophagus.

Host

Human, zoophilic.

Host plant

Grass and other vegetation.

Holotype

Female, India, Maharashtra, Kolhapur, coll. Girhe B.E., 28-IX-2000; head, antenna, wings, legs and abdomen mounted on slide and labelled as above.

Paratype

24 Male, 83 Female Sex ratio (♂ : ♀) 1 : 3.45, coil. Girhe, B.E., September 2000 to September 2001, Kolhapur; 4♂, 12♀, Kolhapur, 24-IX-2000, 3♂, 9♀, Malkapur, 2-X-2000, 2♂, 6♀, Kolhapur, 15-XI-2000, 1♂, 5♀, Kagal, 17-XII-2000, 2♂, 6♀, Panhala, 6-I-2001, 1♂, 7♀, Jaysingpur, 2-II-2001, 3♂, 10♀, Gadhinglaj, 4-III-2001, 1♂, 6♀, Ajra, 5-IV-2001, 1♂, 3♀, Hatkanangale, 4-V-2001, 1♂, 5♀, Kolhapur, 2-VII-2001, 2♂, 8♀, Malkapur, 18-VIII-2001, 3♂, 6♀, Hatkanangale, 3-IX-2001, same data as above.

ANOPHELES (CELLIA) STEPHENSI LISTON

Female (Fig. 26)

4.55 mm long, 1.10 mm broad; antenna 1.35 mm long, hind leg 6.9 mm long, wing 3.2 mm long, head blackish brown, thorax blackish brown, abdomen reddish brown.

Head

0.70 mm long, 0.50 mm broad, dark blackish brown; eyes large, rounded, black; interocular space 0.11 mm; clypeus rounded, brown; proboscis, 1.65 mm long, cylindrical, blackish brown; palpi 1.65 mm long, slender with speckling, as long as proboscis; apical and subapical pale band of palpi equal and separated by a dark band; nape triangular , brown; antenna 1.35 mm long, 15 segmented, hairy brown, pilose; scape 0.22 mm long and 0.15 mm broad; pedicel 0.08 mm long and 0.10 mm broad, brown, rounded; flagellum 1.05 mm long, 13 segmented.

Flagellar formula

1 L/W = 0.12, 13 L/W = 0.02, L 1/13 = 0.085, W 1/13 = 0.19.

Thorax

1.20 mm long, 1.10 mm broad, blackish brown, broad scaled, laterally compressed; anterior pronotal lobe blackish brown, sparesely hairy; scutum black, shield like, covers greater part of thorax dorsally; prealar knob and stemopleuron yellowish brown; prescutellar space yellow, sparesely hairy; scutellum rounded, black, half moon shaped with uniform row of hairs along the margin; anterior pronotal lobe and posterior pronotal lobe brown, pronotum rounded, lack setae; spiracles rounded, yellow.

Fore wing

3.2 mm long, 6.5 mm broad, spotted with white dark scales, scales 0.02 mm long, present on veins as well as fringe on hind margin; subcosta (SC) straight, 2.65 mm long, extending the costa, R_{2+3} forked; media (M) straight 2.37 mm long, Radius straight, slightly curved at apex, 2.90 mm long; cubitus (cu) bifurcated, cu_1 1.43 mm long; anal (A) 1.39 mm long and reaching wing margin, wing with four dark spots on costa involving vein.

Angle between m-cu and cu_1 93°, m_{3+4} and r-m 47° and R_{2+3} and r- m 36°.

Halter

0.13 mm long, 0.10 mm broad and rounded at tip.

Hind leg

6.9 mm long, slender, speckled, longer than body; coxa 0.5 mm long, broad, brown; trochanter 0.20 mm long, hard, black; femur 1.0 mm long, cylindrical; tibia 1.50 mm long, slender with brownish scales arranged densely; tarsus 3.7 mm long and 5 segmented, tarsal segments without bands; claws simple and curved.

Other legs

Special marks: Similar, unbanded.

Abdomen

2.65 mm long, blackish brown, slightly tapered at distal end; tergites blackish brown, while sternite yellowish; abdomen unbanded, densely hairy, scales absent; anal cerci 0.15 mm long, small, yellowish brown, club shaped.

Colour

Black	:	Head, scales on wing.
Blackish brown	:	Thorax and abdomen, tergites.

Brown	:	Nape.
Yellowish brown	:	Anal cerci, pulvilli.
Yellow	:	Mandible, maxillae. White: Scales on wing.

Male

5.00 mm long, slender, smaller than female, antenna plumose; phytophagous.

Host

Human, cattle.

Host plant

Grass, other vegetation.

Holotype

Female, India, Maharashtra, Kolhapur, coll. Girhe, B.E., 28-IX-2000; head, antenna, wings, legs arid abdomen mounted on slide, labelled as above.

Paratype

17 Male, 59 Female, Sex ratio (♂ : ♀) 1: 3.47, coll. Girhe, B.E., September 2000 to September 2001; 8♂, 26♀, Kolhapur, 11-X-2000; 6♂, 21♀, Malkapur, 18-VIII-2001; 3♂, 12♀, Ajra, 3-IX-2001, same data as above.

ANOPHELES (CELLIA) THEOBALDI GILES

Female (Fig. 27)

4.3 mm long, 0.58 mm broad, antenna 1.25 mm long, fore wing 2.65 mm long, hind leg 7.97 mm long, head blackish brown, thorax blackish brown, abdomen yellowish brown.

Head

0.65 mm long, 0.45 mm broad in dorsal view nearly rounded or globular; eyes large, rounded, black, interocular space 0.16 mm; vertex smooth, dark brown; nape rounded, reddish brown; proboscis 1.10 mm long, cylindrical, black, densely scally; palpi slender, as long as proboscis, black banded; apical and subapical pale bands equal and separated by a dark band; mandibles maxillae yellow, well developed, toothed; antenna 1.25 mm long, 15 segmented, hairy, pilose; scape 0.15 mm long, 0.13 mm broad; pedicel 0.25 mm long, 0.09 mm broad, blackish brown; flagellum 0.85 mm long, 13 segmented.

Flagellar formula

1 L/W = 0.13, 13 L/W = 0.21, L 1/13 = 0.12, W 1/13 = 0.07.

Thorax

1.10 mm long and 0.75 mm broad, blackish brown, laterally compressed, lyre shaped, anterodorsocentral, posterodorsocentral bristles present; stemopleuron broad, yellowish brown; postnotum smooth, yellowish brown, rounded; prescutellar space brownish, sparesely hairy; scutellum half moon shaped with a uniform row of hairs along the margin; pronotum lack setae.

Fore wing

2.65 mm long and 0.7 mm broad, spotted with white and dark scales, scales 0.02 mm long, present on veins as well as fringe on hind margin, subcosta (SC) straight, 2.00 mm long, reaching the costa, R_{2+3} forked, R_{4+5} simple, no cross vein connection of R_1 and R_2, media (M) straight, 1.67 mm long, then bifurcated; radius straight, slightly curved at apex, 2.45 mm long; cubitus bifurcated, cu_1 1.8 mm long; anal vein 1.6 mm long, wing with four dark spots on costa including vein 1, fringe spot on all the veins.

Angle between m-cu and cu_1 87°, m_{3+4} and r-m 75° and R_{2+3} and r-m 31°.

Halter

0.20 mm long, brown and 0.18 mm broad and rounded at tip (pin shaped).

Hind leg

7.95 mm long, slender, longer than body; coxa 1.10 mm long, broad, brownish yellow; trochanter 0.3 mm long, hard, yellowish brown; femur 1.85 mm long, speckled, cylindrical; tibia 2.10 mm long, speckled; tarsus 2.60 mm long, 5 segmented, slender, tarsal segments banded, apicel white band of 3rd, 4th and 5th segment white; claws simple, curved.

Other legs

Special marks: tarsal segments on fore leg with broad bands, mid leg similar.

Abdomen

2.55 mm long, 0.55 mm broad; bands absent, densely hairy, dorsally yellowish brown, ventrolaterally blackish brown; scales absent; anal cerci very small, hairy, 0.14 mm long, 0.11 mm broad.

Colour

Black	:	Eye, proboscis.
Blackish brown	:	Scape, pedicel.
Brown	:	Vertex, tarsal segments.
Yellow	:	Mandibles, maxillae.
Yellowish brown	:	Sternopleuron, postnotum.
White	:	Bands on tarsal segments.

Male

5.00 mm long, slender, smaller than female, antenna plumose, phytophagous.

Host

Human, domestic animals.

Host plant

Grass and other vegetation.

Holotype

Female, India, Maharashtra, Kolhapur, coll. Girhe, B.E. 19-VII-2001; head, antenna, leg, wing and abdomen mounted on the slide labelled as above.

Paratype

8 Male, 18 Female, coll. Girhe, Sex ratio (♂ : ♀) 1 : 2.25, September 2000 to September 2001, Kolhapur; 20♂, 5♀, Gadhinglaj 17-VI-2001; 40♂, 8♀, Malkapur, 11-VII-2001; 20♂, 5♀, Hatkanangale, 3-IX-2001, same data as above.

ANOPHELES (CELLIA) SUBPICTUS GRASSI

Female (Fig. 28)

5.20 mm long, 1.25 mm broad, fore wing 3.00 mm long, antenna 1.30 mm long, hind leg 6.8 mm long, head brown, thorax and abdomen blackish brown.

Head

0.40 mm long, 0.37 mm broad, brown, hairy, in dorsal view nearly rounded or globular; clypeus blackish brown, rounded; vertex yellowish brown with vertical bristles; tempus semicircular, black with temporal bristles; eyes black, rounded, large, inter ocular space 0.018 mm; proboscis 1.62 mm long, blackish brown; labellum yellow, small; palpi 1.62 mm long, as long as proboscis; apical white pale band nearly equal to the pre-apical dark band; mandibles and maxillae slender, as long as proboscis, toothed; nape rounded, black with flat scales; antenna 1.30 mm long, brown, pilose, 15 segmented; scape 0.12 mm long, 0.18 mm broad; pedicel 0.09 mm long, 0.15 mm broad, rounded and blackish brown; flagellum 1.09 mm long, 13 segmented.

Flagellar formula

1 L/W = 0.25,13 L/W = 0.61. L 1/13 = 0.7, W 1/13 = 0.17.

Thorax

1.55 mm long, 0.85 mm broad, scutum blackish brown, covers greater part of thorax dorsally, scutellum half moon shaped with uniform row of hairs along the margin; mesothoracic spiracle rounded and with bristles; postnotum smooth and blackish brown; pre-scutellar space brownish and sparesely hairy; anterodorsocentral bristles and posterodorsocentral bristles present; anterior and posterior pronotal lobes rounded, black; pronotum lack setae.

Fore wing

3.00 mm long, 0.85 mm broad with white and dark scales; scales 0.02 mm long, present on veins as well as fringe on hind margin; wing with four dark spots on costa vein 1; fringe spot on all the veins, Sc vein straight, 2.60 mm long, reaching the costa, R_{2+3} forked; media straight, bifurcated apically, 2.30 mm long, radius straight, slightly curved at apex, 2.90 mm long, cross-vein r-m and m-cu both present; cubitus (cu) bifurcated, cu_1 1.68 mm long; anal vein 1.42 mm long, touching wing margin.

Angle between m-cu and cu_1 95°, m_{3+4} and r-m 83° and R_{2+3} and r- m 33°.

Halter

0.13 mm long, 0.08 mm broad, blackish brown, rounded at tip.

Hind leg

7.8 mm long, yellowish brown, elongated than body; coxa 0.7 mm long, yellowish, broad, brown; trochanter 0.2 mm long, dark brown; femur 2.00 mm long, without spur, black, non speckled; tibia 2.30 mm long, yellow, elongated than femur, non speckled; tarsus 2.6 mm long, five segmented with narrow five white bands on each segment; claws simple and curved.

Other legs

Special marks: Similar.

Abdomen

3.25 mm long, 0.6 mm broad, blackish brown, not distinctly banded, densely hairy, scales absent, anal cerci 0.25 mm long, 0.22 mm broad, brown.

Colour

Black	:	Eye, scales on wing.
Blackish brown	:	Scutum, postnotum.
Brown	:	Proboscis, antenna.
Yellowish brown	:	Hind leg.
Yellow	:	Mandible, maxillae, labellum.
White	:	Bands on tarsus of each leg.

Male

4.90 mm long, slender, smaller than female, antenna plumose, phytophagous.

Host

Human and domestic animals.

Host plant

Vegetation.

Holotype

Female, India, Maharashtra, Kolhapur coll. Girhe, B.E., 16-XI-2000, head, antenna, leg, wing and abdomen mounted on the slide labelled as above.

Paratype

6 Male, 19 Female, Sex ratio (♂ : ♀) 1: 3.16, coll. Girhe, B.E., September 2000 to September 2001, Kolhapur; 1♂, 3♀, Kolhapur 29-IX-2000; 1♂, 4♀, Kolhapur, 15-XI-2000; 4♂, 12♀, Malkapur 11-VIII-2001, same data as above.

GENUS *CULEX* LINNAEUS 1758

The genus *Culex* is erected by Linnaeus in 1758. More than 400 species of this genus have been reported from various parts of the world (Stone *et al.*, 1959). It is characterized by:

1. Presence of well developed pulvilli,

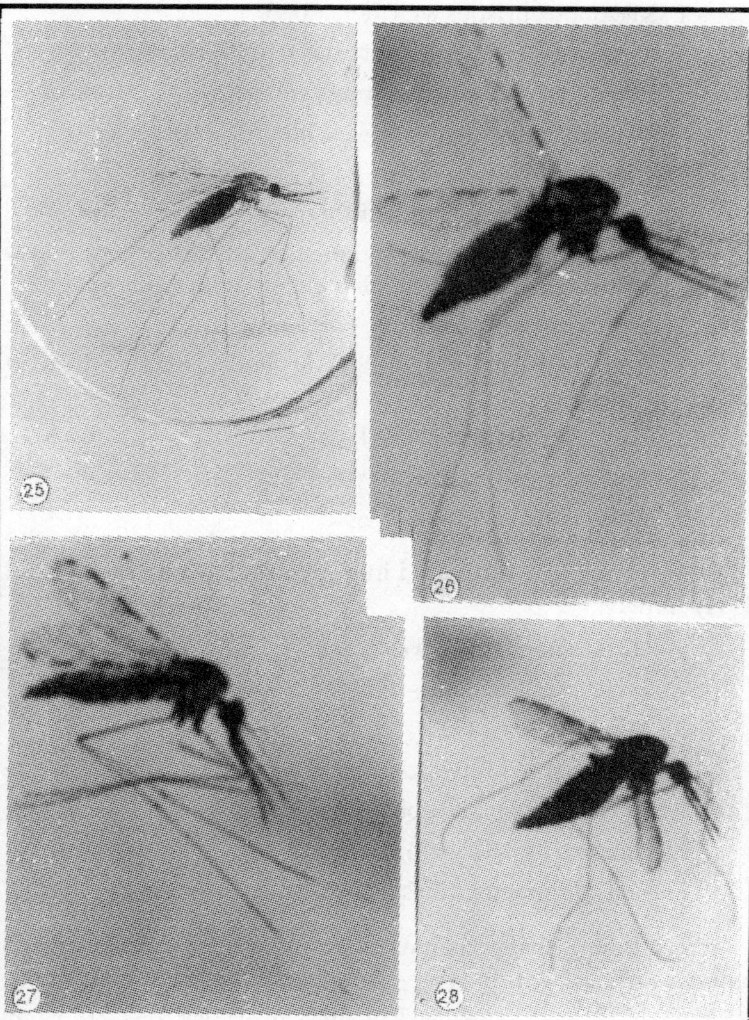

Plate III

Fig. 25: ***Anopheles culicifacies*****, Fig. 26:** ***Anopheles stephensi*****, Fig. 27:** ***Anopheles theobaldi*****, Fig. 28:** ***Anopheles subpictus*****.**

2. Buccopharyngeal armature present in female,
3. No spircular or post spiracle bristles,
4. Mesonotal bristles well developed,
5. Mesonotal and scutellar scales usually all narrow,
6. In female, palpi shorter than proboscis,
7. Wings generally unspotted with only dark scales,
8. Tarsi without pale ring,
9. Scutellum is trilobed with three tuft of hairs on the lobe,
10. Palpi four segmented, may be entirely dark, tipped or ringed with white and some times covered with white spot,
11. Pronotum lack setae,
12. Abdomen slightly tapering, blunt and covered with broad scales,
13. Claws simple and curved,
14. Abdomen covered with scales and
15. Anal cerci very small.

The genus *Culex* is divided into seven subgenera by Barraud (1934) namely:

1. *Lutzia* Theobald, 1903,
2. *Barraudius* Edwards, 1921,
3. *Neoculex* Dyar, 1905,
4. *Mochthogenes* Edwards, 1930,
5. *Lophoceratomyia* Theobald, 1905,
6. *Culiciomyia* Theobald, 1907 and
7. *Culex* Linnaeus, 1758.

From India, 50 species have been reported under above sub genera. *Lutzia* contain 4 species, *Barraudius* 1, *Neoculex* 2, *Mochthogenes* 5, *Lophoceratomyia* 10 and *Culiciomyia* 7 species while sub genus *Culex* contain 21 species.

SUB-GENUS *CULEX*

Syst. Nat. ed. X, p. 602. Genotype, *C. pipiens* Linn.

Leucomyia Theobald, 1907, M.C. iv, p. 372.

Genotype, *C. gelidus* Theo.

Oculeomyia Theobald, 1907, M.C. iv, p. 515.

Genotype *C. bitaeniorhynchus* Giles (as 0. *sarawati* Theo.)

Theobaldiomyia Brunetti, 1912 Ree. Ind. Mus, iv p. 462 (nom. nov for *Leucomyia* Theo.) (Edwards, 1932, p. 200).

The subgenus *Culex* is characterized by:

1. Moderate body size,
2. Head with upright and narrow scales on vertex, flat scales at sides but no border of flat scales round eye margin infront,
3. Scutellar scales narrow,
4. Palpi of male always longer than proboscis, last two segments up curved and hairy,
5. Antenna of male without scale tufts,
6. Hypopysium without scales and without an apicoventral lobe but with a subapical lobe divided,
7. Style without spiny crest,
8. Phalosome divided into lateral portions usually with various teeth or lobes,
9. Paraprocts with rather dense tuft of spines or hairs at crown and
10. Buccophatyngeal armature of female with one row of teeth, teeth may be sharp or blunt.

From Kolhapur district two species have been reported.

CULEX (CULEX) EPIDESMUS THEOBALD

Female (Fig. 29)

8.65 mm long, 2.00 mm broad, antenna 2.9 mm long, fore wing 5.25 mm long, hind leg 14.4 mm long, head yellowish brown, thorax reddish brown, abdomen yellowish brown.

Head

1.2 mm long, 0.65 mm broad, globular with upright and narrow scales on vertex, yellowish brown, flat scale at sites but no border, flat scale around eye margin in front; clypeus broad, rounded, convex dorsally; eyes black, interocular space 0.12 mm; tempus semi-circular, dark brown; proboscis 3.00 mm long, brown with pale ring in the middle, cylindrical and straight; labium cylindrical, densely scally and dark brown; labellum short and pale brown; maxillary palp four segmented, dark brown, 0.6 mm long, stub like, hairy, shorter than proboscis; mandibles and maxillae long, blade like; stylet curved; palpifer and palpus brown; nape dark brown and rectangular; antenna 2.9 mm long, 15 segmented, blackish brown and pilose; scape 0.05 mm long, 0.18 mm broad, dark brown, pedicel 0.10 mm long, 0.15 mm broad; flagellum 2.75mm long, 13 segmented.

Flagellar formula

1 L/W = 1.5,13 L/W = 0.62, L 1/13 = 0.03, W 1/13 = 0.12.

Thorax

1.75 mm long, 1.6 mm broad, reddish brown, undifferentiated, laterally compressed; scutum blackish brown, shield like, covers greater part of the dorsal surface; sternopleuron rectangular, reddish brown and larger than mesepimeron; anterior pronotallobe rounded, brown; mesepimeral bristles two; paraprocts with dense tuft of spines; prescutellar space brown with prescutellar bristles; scutellum brown, trilobed with three tufts of hair on the lobe; scutellar scales narrow; anterodorsocentral bristles and posterodorsocentral bristles blackish brown.

Fore wing

5.25 mm long, 1.40 mm broad, yellowish area at tip, covered with dark scales; scales 0.02 mm long, present on veins as well as fringe on hind margin; costa dark except at tip; subcosta straight 4.63 mm long, R_{4+5} simple, no cross-vein connection of R_1 and R_2; media (M) straight, branched apically, 3.3 mm long; radius 5.15 mm long, straight, slightly curved at apex; cross-vein r-m and m-cu both present; cubitus (CU) forked, anal (A) 1.55 mm long and reaching wing margin.

Angle between m-cu and cu_1 89°, m_{3+4} and r-m 81° and R_{2+3} and r-m 35°.

Halter

0.65 mm long, 0.42 mm broad, rounded, expanded at tip.

Hind leg

14.4 mm long, elongated, brown, slender longer than body; coxa 1.3 mm long, brown, yellowish brown with coxal bristles; trochanter

0.6 mm long, hard, brown; femur 3.5 mm long, cylindrical, densely covered with scales and spurs; tibia 4.2 mm long, slender, straight with yellowish brown tibial spurs; tibia longer than femur; tarsus 4.8 mm long, with pale ring, five segmented; first pretarsus longer than other; claws simple, curved, brownish, untoothed; pulvillus and empodium densely hairy.

Other legs

Special marks: Similar.

Abdomen

5.70 mm long, 1.2 mm broad with dark yellow scales; dorsal abdominal pale stripes parallel, unconstricted dark scales rare on underside and restricted to centre line; last abdominal segment narrow and shorter than other; anal cerci 0.4 mm long. 0.3 mm broad, brown.

Colour

Black	:	Eyes.
Blackish brown	:	Scutum, vertex, scape, antenna
Brown	:	Proboscis, legs, maxillary palp, Scutellum.
Yellow	:	Wing, femur, coxa.
Brown	:	Scutum, sternopleuron.

Male

8.50 mm long, smaller, slender than female, antenna elongated, plumose and brushy; phytophagous.

Host

Human.

Host plant

Unknown.

Holotype

Female, India, Maharashtra, Kolhapur. coll. Girhe, B.E., 19-X-2000; Head, antenna, leg, abdomen, mounted on slide, labelled as above.

Paratype

4 male, 29 female; Sex ratio (M:F) 1 : 7.25; coll. Girhe, B.E., from September 2000 to September 2001, Kolhapur; 1♂, 9♀. Kolhapur, 21-

XII-2000; 2♂, 15♀, Hatkanangale, 6-I-2001; 1♂, 5♀, Gadhinglaj, 4-III-2001; same data as above.

CULEX (CULEX) PIPIENS LINNAEUS

Female (Fig. 30)

7.0 mm long, 1.2 mm broad, antenna 2.75 mm long, brownish, hairy; fore wing 4.5 mm long, hind leg 10.4 mm long; head, thorax and abdomen brown.

Head

1.20 mm long, 1.0 mm broad, globular and nearly rounded; vertex rough, brownish laterally hairy; eyes large, rounded, black; interocular space 0.45 mm; clypeus rounded, brown, nape rectangular, brown; vertical bristles and temporal bristles elongated and yellowish brown; proboscis 2.2 mm long, with pale ring in the middle, brown, straight, club shaped; palpi 0.6 mm long, dark brown, four segmented, densely hairy, stub like; mandibles and maxillae long, straight, yellow with blade like stylets; antenna 2.75 mm long, 15 segmented, dark brown, hairy , pilose; scape 0.08 mm long, 0.22 mm broad; pedicel 0.12 mm long, 0.19 mm broad, dark brown; flagellum 2.55 mm long, 13 segmented.

Flagellar formula

1 L/W = 0.13, 13 L/W = 0.32, L1/ 13 = 0.063, W 1/13 = 0.15.

Thorax

1.5 mm long, 2.1 mm broad, laterally compressed, undifferentiated, brown; scutum shield like, dark brown, covers dorsal surface; anterior prenotal lobe nearly rounded, brown; spiracular area pale yellow; mesepimeron rectangular , yellow two bristles present; scutellum yellowish brown, trilobed with three bunches of hair on lobe; metapleuron triangular; prealar knob rounded, yellow; paratergite elongated and pale yellow.

Fore wing

4.5 mm long, 1.10 mm broad, unspotted with only dark scales, scales 0.02 mm long, present on veins as well as fringe on hind margin; subcosta 3.70 mm long, straight, reaching the costa, base of subcosta without tuft of hairs on the under side of the wing, media (M) straight, 3.84 mm long, radius straight slightly curved at apex, 4.3 mm long, r-m and m-cu present; cubitus (cu) bifurcated, cu_1 1.52 mm long; anal (A) vein 1.14 mm long, extend up to wing margin.

Angle between m-cu and cu_1 91°, m_{3+4} and r-m 76° and R_{2+3} and r-m 36°.

Halter

0.45 mm long, 0.26 mm broad, yellowish brown, club shaped at tip.

Hind leg

10.4 mm long, slender, longer than body; coxa 1.0 mm long, yellowish brown, broad, with few brownish bristles; trochanter 0.3 mm long, hard, dark brown; femur 2.5 mm long, cylindrical; preapical end pale yellow while apical end yellowish brown, with distinct white spot; tibia 3.0 mm long, slender with brownish tibial spines; tarsus 3.6 mm long, five segmented, without pale ring; claws simple, curved, untoothed.

Other legs

Special marks: Unbanded, dark brown, pale ring absent.

Abdomen

4.3 mm long, 1.4 mm broad, elongated, slender, covers with thickly overlapping scales; tergites dark brown while sternites pale yellow; dorsal abdominal pale stripes constricted laterally and centrally, patches of dark scales on under side of pale abdomen on median line; abdomen covers with few hairs, last two segments densely hairy; anal cerci 0.2 mm long, 0.18 mm broad, dark brown, densely hairy, broad.

Colour

Black	:	Eyes.
Brownish black	:	Head.
Brown	:	Clypeus, tergites, legs, claws, palpi.
Yellow	:	Mandibles and maxillae, sternites.
Yellowish brown	:	Scutellum, femur.

Male

6.9 mm long, smaller and slender than female, antenna plumose and brushy; phytophagous.

Host

Human and birds.

Host plant

Grass and other vegetation.

Holotype

Female, India, Maharashtra, Kolhapur. coll. Girhe B.E., 11-1-2001; Head, antenna, leg, abdomen mounted on slide, labelled as above.

Paratype

15 male, 43 female; Sex ratio (M:F) 1 : 2.86; coll. Girhe, B.E., from September 2000 to September 2001, Kolhapur; 1♂, 4♀, Kolhapur, 29-IX-2000; 2♂, 6♀, Malkapur, 17-X-2000; 1♂, 5♀, Kolhapur, 30-XI-2000; 2♂, 3♀, Kagal, 3-XII-2000; 1♂, 2♀, Panhala, 6-I-2001; 1♂, 3♀, Jaysingpur, 17-II-2001; 1♂, 2♀, Gadhinglaj, 4-III-2001; 1♂, 3♀, Ajra, 5-IV-2001; 1♂, 4♀, Hatkanangale, 4-V-2001; 1♂, 5♀, Kolhapur, 28-VII-2001; 1♂, 3♀, Malkapur, 31-VIII-2001; 2♂, 3♀, Hatkanangale, 2-IX-2001; same data as above.

SUB-GENUS *BARRAUDUS* EDWARDS

Bull. Ent. Res., xii, p. 322. Genotype, c. *pusillus* Maeq.

The subgenus *Barraudus* erected by Edwards in 1921 is distinguished by following characters:

1. Small mosquitoes without conspicuous ornamentation.
2. Segment one of hind tarsi distinctly shorter than tibia.
3. Pulvilli well developed.
4. Scales on vertex of head, apn and scutellum all narrow.
5. Paraproct with hairs or spines at crown, no lateral arms.

The subgenus *Barraudus* is represented by a single species from India (Barraud, 1934).

CULEX (BARRUDUS) MODESTUS FICALBI

Female (Fig. 31)

7.30 mm long, 1.12 mm broad, fore wing 4.5 mm long, antenna 2.3 mm long, hind leg 13.0 mm long, head dark brown, thorax yellowish brown, abdomen yellow.

Head

0.80 mm long, 0.72 mm broad, covered with brown narrow scales, globular dark brown; clypeus broad, rounded, dark brown; tempus yellow, vertex pale yellow; vertical and temporal bristles elongated, dark brown; tori yellowish brown; eyes black, rounded, interocular space 0.025 mm; maxillary palp four segmented; palpi 0.72 mm long, dark brown, stub like, hairy, ¼ of proboscis; proboscis 2.4 mm long, labium yellow; nape rounded with fairly numerous upright scales; antenna 2.3 mm long, 15 segmented, brownish, pilose and hairy; scape 0.18 mm long, 0.25 mm broad, rounded, yellow; pedicel 0.20 mm long and 0.21 mm broad; flagellum 1.92 mm long, 13 segmented.

Flagellar formula

1 L/W = 0.25, 13 L/W = 0.63, L 1/13 = 0.08, W 1/13 = 0.20.

Thorax

2.00 mm long, 1.10 mm broad, laterally compressed with hump shaped appearance; scutum shield like, yellowish brown; scutellar scales lighter, anteriopronotal lobe semicircular, dark brown, posterior pronotal lobe flat and yellow; scutellum pale yellow, trilobed with three bunches of hair on the lobe; sternopleuron and mesepimeron large, flat and yellowish brown; prespiracular area pale yellow; anteriodorsocentral bristles and posteriodorsocentral bristles elongated, yellowish brown.

Fore wing

4.50 mm long, 0.90 mm broad, unspotted with only dark scales, scales 0.019 mm long, present on veins as well as fringe on hind margin; subcosta (SC) vein 3.72 mm long, straight, reaching the costa; R_{2+3} forked, R_{4+5} simple, media (M) 3.96 mm long, straight, bifurcated apically; cross-vein r-m and m-cu both present; cubitus (cu) bifurcated, cu_1 3.38 mm long; radius straight and slightly curved at appex, 4.30 mm long; anal (A) vein 2.65 mm long, reaching margin.

Angle between m-cu and cu_1 92°, m_{3+4} and r-m 73° and R_{2+3} and r-m 39°.

Halter

0.30 mm long, 0.19 mm broad, pale yellow with broad tip.

Hind leg

13.00 mm long, yellowish brown, longer than body; coxa 0.60 mm long, broad, hairy and yellowish brown; trochanter 0.20 mm long, black and hard; femur 2.0 mm long, cylindrical; tibia 3.5 mm long, densely scally; tarsus 6.7 mm long, 5 segmented, unbanded; segment 1 of hind tarsus shorter than tibia; claws simple and curved; pulvillus and empodium narrow.

Other legs

Special marks: All legs dark brown, fore and hind femora lighter posteriorly, pale except dorsally.

Abdomen

4.5 mm long, 1.10 mm broad; dorsum dark brown, lateral rectangular pale ochreous patches on tergites forming a continuos pale border on either side of abdomen, sternites uniformly pale ochreous; abdomen cylindrical, sparsely hairy, dark yellow dorsally and pale yellow ventrally. Abdomen not distinctly banded, brownish scales restricted on the tergites. Anal cerci 0.15 mm long and 0.12 mm broad.

Colour

Black	:	Eye, trochanter.
Dark Brown	:	Proboscis, flagellum of antenna, clypeus, Palpi.
Yellow	:	Sternites, mandibles.
Yellowish brown	:	Abdomen.

Male

7.00 mm long, slender, smaller than female, antenna plumose and brushy, phytophagous.

Host

Human, cattle.

Holotype

Female, India, Maharashtra, Kolhapur, coll. Girhe, B.E., 7-VI-2001; head, antenna, legs, abdoment, mounted on slide, labelled as above.

Paratype

6 male, 21 female; Sex ration (M:F), 1:3.5; coll. Girhe, B.E. September 2000 to September 2001; 1♂, 5♀, Kolhapur, 29-IX-2000; 3♂, 11♀, coll. Sathe, T.V., Kolhapur, 27-VI-2001; 2♂, 5♀, coll. Girhe, B.E., Hatkanangale, 2-IX-2001; as data as above.

SUBGENUS *MOCHTHOGENES* EDWARDS

Bull. Ent. Res. XXI, p.305. Genotype, *C. malayi* (Leic).

The subgenus *Mochthogenes* is characterized by:

1. Small size and dark colour without ornamentation,
2. Dorsal surface of head mainly covered with flat scales,
3. First hind tarsal segment about as long as tibia or longer and,
4. Palpi quite short in both sexes.

From India five species have been reported (Barraud, 1934).

CULEX (MOCHTHOGENES) MALAYI EDWARDS

Female (Fig. 32)

8.00 mm long, 1.10 mm broad, antenna 2.95 mm long, fore wing 4.6 mm long, hind leg 9.3 mm long; head blackish brown, thorax blackish brown, abdomen pale yellow.

Head

1.00 mm long, 0.8 mm broad, covered with small light coloured flat scales; in dorsal view nearly rounded or globular, blackish brown; eyes black, large; interocular space 0.15 mm, clypeus rounded, brown; vertex yellowish brown, with vertical bristles; tempus semicircular brown with temporal bristles; upright scales an vertex and nape; proboscis 1.40 mm long, straight, cylindrical, brownish black; labellum small, yellow; palpi 0.4 mm long, hairy, four segmented, 1/6th length of proboscis, entirely brownish black; mandibles and maxillae, slender, straight, yellow with blade like stylet, as long as proboscis; buccopharyngeal armature present; head with flat scales at side and with a small flat scales along the eye margin; antenna 2.95 mm long, hairy, 15 segmented, pilose, brownish black, longer than proboscis; scape 0.08 mm long, 0.18 mm broad; pedicel 0.11 mm long, 0.14 mm broad, rounded and brown; flagellum 2.76 mm long, 13 segmented.

Flagellar formula

1 L/W = 0.15,13 L/W = 0.68, L 1/13 = 0.03, W 1/13 = 0.13.

Thorax

2.25 mm long, 1.15 mm broad, laterally compressed, blackish brown, shield like, scutum covers dorsal surface; anterior pronotal lobe and posterior pronotal lobe rounded; mesonotum with brown narrow scales; scutellar scales narrow, acrostichal bristles elongated; fossal area yellow; scutellum trilobed with three tuft of hairs on the lobe; metapleuron formed by metapisternum, metepimeron and metameron; mesothoracic spiracle rounded while metathoracic spiracle slit shaped; anterior dorsocentral bristles and posterior dorsocentral bristles blackish brown.

Fore wing

4.80 mm long, 1.3 mm broad, unspotted, covered with only dark scales; scales 0.023 mm long, present on veins as well as fringe on margin; subcosta (SC) vein straight, 3.25 mm long, reaching the costa; radius straight, 4.59 mm long; R_{2+3} forked, R_{4+5} simple, no cross-vein connection of R_1 and R_2, cross vein r-m and m-cu prominant; media (M) straight, 3.47 mm long, bilobed apically, cubitus (cu) bifurcated, Cu_1 2.89 mm long; anal 1.50 mm long, reaching wing margin.

Angle between m-cu and cu_1 90°, m_{3+4} and r-m 75° and R_{2+3} and r-m 42°.

Halter

0.40 mm long, 0.22 mm broad, tip club shaped.

Hind leg

9.3 mm long, slender, longer than body; coxa 1.0 mm long, narrow, broad, deep brown; trochanter 0.3 mm long, hard, dark; femur 2.0 mm long, pale narrowly, brown dorsally from near base to knee; cylindrical, preapical region white, dark at apical end; tibia 2.8 mm long, slender; tarsus 3.2 mm long, 5 segmented, with dark scales, without bands; first pretarsal longer than others, claws simple, curved.

Other legs

Special marks: Fore and mid femora lighter posteriorly.

Abdomen

4.75 mm long, 1.2 mm broad, cylindrical, brownish black dorsally; pleuron with several broad scales, each group usually consists of scales and marked with pale bands; abdomen with uniform transverse pale yellow bands and dark scale band alternately; tergites sparesely covered with hairs, sternal plates dirty white with few scales with pale broad yellow band; anal cerci 0.13 mm long, 0.08 mm broad, small, few hairs lies on last two segment.

Colour

Black	:	Eye, abdominal dark band, head.
Reddish brown	:	Scutum, prenotallobe, post notal lobe.
Brownish black	:	Antenna, proboscis.

Yellow	:	Mandibles and maxillae.
White	:	Sternal plates.

Male (Fig. 33)

7.95 mm long, slender and smaller than female, antenna plumose and brushy, phytophagous.

Host

Human.

Host plant

Vegetation.

Holotype

Female, India, Maharashtra, Kolhapur coll. Girhe, B.E., 12-II-2001; Head, antenna, leg and abdomen, mounted on slide, labelled as above.

Paratype

5 Male, 14 Female, Sex ratio (M:F) 1 : 2.8, coll. Girhe, B.E. September 2000 to September 2001, Kolhapur; coll. Sathe, T.V., 2♂, 6♀, Kolhapur , 2-II-2001, coll. Girhe, B.E. 2♂, 4♀, Kagal 5-III-2001, 1♂, 4♀, Ajra, 5-IV-2001; same data as above.

GENUS *AEDES* MEIGEN, 1818

The genus *Aedes* is erected by Meigen in 1818. This genus includes 600 species represented from all over the world (Cheng, 1964) of which more than 112 species have been reported from India (Barraud, 1934). This genus shows following features:

1. Claws toothed in most female species.
2. Post spiracular bristles present.
3. Pulvilli absent or hair like.
4. Female abdomen tends to be more pointed.
5. Cerci longer than in other groups.
6. Fringe of hair on squama of wing.
7. Proboscis normal and not enlarged or approximated behind the head.
8. Antenna of male nearly always distinctly plumose with the last two segments elongate and in female with moderate long hairs.

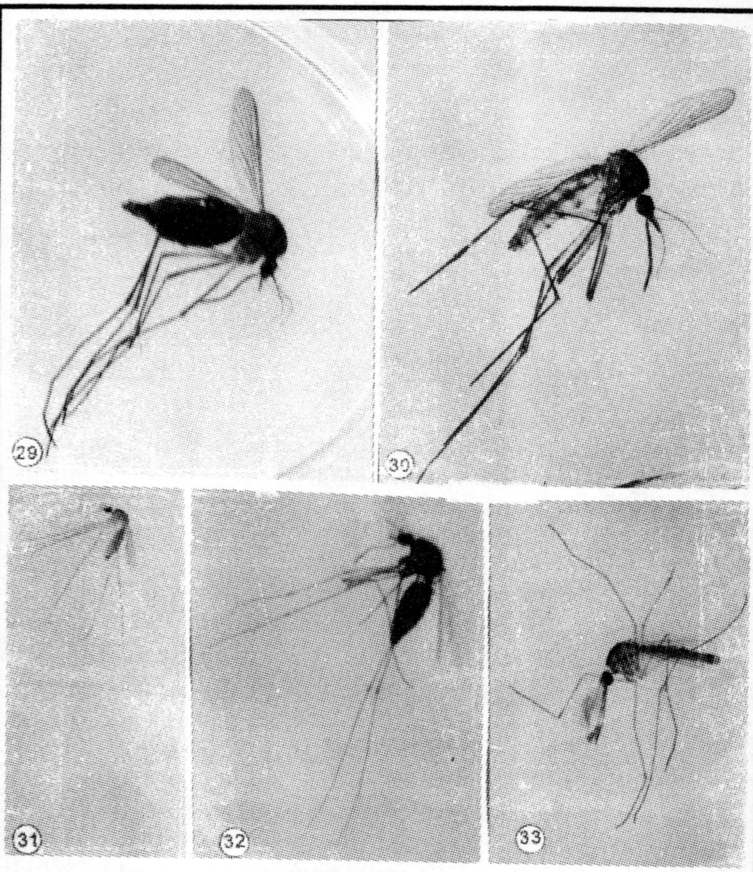

Plate IV

Fig. 29: *Culex epidesmus*, Fig. 30: *Culex pipiens*, Fig. 31: *Culex modestus*, Fig. 32: *Culex malayi* female, Fig. 33: *Culex malayi* male.

9. Bristles of mesonotum usually well developed.
10. Pronotum lack setae.
11. Scutellum trilobed.

Barraud (1934) divided the genus *Aedes* into following 11 subgenera:

1. *Medicus* Theobald, 1901,
2. *Ochlerotatus* Lynch Arribalzaga, 1891,
3. *Finlaya* Theobald, 1903,
4. *Christophersiomyia* Barraud, 1923,
5. *Stegomyia* Theobald, 1901,
6. *Rhinoskusea* Edwards, 1929,
7. *Indusius* Barrauds, 1934,
8. *Bankinella* Theobald, 1907,
9. *Aedes* Meigen, 1818,
10. *Cancredes* Edwards, 1929, and
11. *Diceromyia* Theobald, 1911.

From India, Barraud (1934) described 116 species belonging to above eleven subgenera.

SUBGENUS *STEGOMYIA* THEOBALD

M.C. I, p. 233. Genotype, *Culex fasciatus* Fab.

Scutomyia Theobald, 1904, Entom. XXXVii, p. 77.

Genotype S. *albolineata* Theo.

Quasistegomyia Theobald, 1906, 2nd Rept. Wellac. Lab. P. 69.

Genotype Q. *unilineata* Theo.

Catatassomyia Dyar and Shannon, 1925, Insec. Ins. Mens. Xiii, p. 71,

Genotype, C. *neronephada* D. and S.

The subgenus *Stegomyia* is represented by the following features:

1. Body size small or medium.
2. Body colour black or dark, highly ornamented with patches, spots or lines of snow white scales.
3. Two or more basal white bands or tarsi of at least one pair of leg of which one or more tarsal segments completely white. Tarsi in no case all completely dark nor with apical and basal bands.
4. Proboscis entirely dark except in few cases.
5. Scales on vertex and scutellum are broad and flat.

6. Palpi of male slender, terminal segments unruptured and with few hairs.
7. Antennae of male with plumose hairs mainly projecting on two side and not in regular whorles.
8. Harpago absent.

Under this subgenus, Barraud described 16 species from India. From Kolhapur district seven species have been reported under the subgenus *Stegomyia* of which two are redescribed and four species have been newly reported and described for the first time from India.

AEDES (STEGOMYIA) AEGYPTI LINNAEUS

Female (Fig. 34)

5.30 mm long, 0.8 mm broad; antenna 1.95 mm long, fore wing 3.5 mm long, hind leg 8.5 mm long; head blackish brown, thorax black, abdomen dark.

Head

0.6 mm long, 0.5 mm broad, in dorsal view nearly rounded or globular; with silver white flat scales in middle of vertex, continued downward between eyes; silvery flat scales on tori, eyes large, rounded, black; interocular space 0.09 mm; vertex hairy, dark black with vertical bristles; tempus semicircular, blackish brown with temporal bristles; clypeus rounded with two small silvery white dots; scales on vertex flat, broad; proboscis 2.2 mm long, straight, cylindrical, black; palpi 0.6 mm long, shorter than proboscis, palpifer black, palpus white; silver scales on clypeus and palpi; nape rectangular; mandibles, maxillae as long as proboscis, yellow straight, with blade like stylet; antenna 1.95 mm long, blackish brown, 15 segmented, pilose, hairy; scape 0.08 mm long, 0.14 mm broad; pedicel 0.20 mm long, 0.11 mm broad; rounded, blackish brown; flagellum 1.67 mm long, 13 segmented.

Flageller formula

1 L/W = 0.091, 13 L/W = 0.66, L 1/13 = 0.013, W 1/13 = 0.10.

Thorax

1.0 mm long, 0.8 mm broad flat, silver white scales on all lobes of scutellum; laterally compressed, scutum black with scutal suture; white mid dorsal strip present in mid dorsal region of scutum; sternopleuron and mesepimeron broad, flat with patches of white scales; post spiracle rounded with bristles; scales on scutum broad and flat, prescutellar space hairy, black with prescutellar bristles; pronotum lack setae; bristles of mesonotum usually well developed; with lyre shaped marking of silvery scales; anterodorsocentral and posterodorsocentral bristles present.

Fore wing

3.8 mm long, 0.95 mm broad, unspotted only with dark scales; squama and alula convex, transparent, fringe on hind margin; scales present along the veins as well as fringe of them along hind margin; subcosta (SC) straight, 3.2 mm long, reaching the costa, R_{2+3} forked, Media (M) straight but, bifurcated apically, 3.4 mm long, radius 3.7 mm long, straight and slightly curved at end; no cross vein connection of R_1 and R_2, corss-vein r-m and m-cu distinct, cubitus (cu) bifurcated, cu_1 1.62 mm long; anal vein 1.36 mm long, reaching margin.

Angle between m-cu and cu_1 91°, m_{3+4} and r-m 71° and R_{2+3} and r-m 35°.

Halter

0.18 mm long, 0.14 mm broad, yellow, rounded at tip.

Hind leg

8.5 mm long, slender, longer than body; coxa 1.10 mm long, broad, dark-brown, with few bristles; trochanter 0.4 mm long, hard, dark brown; femur 1.4 mm long, cylindrical with a white longitudinal line running from the base for nearly whole length; preapical end white, apical end black with white knee spot; tibia 2.20 mm long, dark, black; tarsus 3.4 mm long, five segmented with prominent five dark and white bands, on 4th usually widest band, 5th segment entirely white; claws curved, toothed; pulvillus short, hairy.

Other legs

Special marks: A narrow basal white band on 1st, 2nd and 3rd tarsal segment of fore and mid leg.

Abdomen

3.7 mm long, 0.7 mm broad, tergites brownish black, slender, pointed; black with dorsal white transverse bands; white patches of scales on latero tergites; sternal plates dirty white, tergal plates dark blackish brown; all abdominal segments clothed with dark and white scales; anal cerci 0.3 mm long, black, flat, sparesely hairy.

Colour

Black	:	Eyes, proboscis, prescutellum.
Blackish brown	:	Abdomen, antenna, tergites.
White	:	Patches on head, thorax and abdomen.
Yellow	:	Mandible, maxillae.
Yellowish brown	:	Wings.

Male

5.20 mm long, slender, smaller than female, antenna plumose, brushy, last two segments elongate; phytophagous.

Host

Human, cattle.

Host plant

Grass and other vegetation.

Holotype

Female, India, Maharashtra, Kolhapur coll. Girhe, B.E., 20-IV-2001, Mouth parts antenna, wing, leg, abdomen on slide, labelled as above.

Paratype

6 Male, 30 Female; Sex ratio (♂ : ♀) = 1 : 5, coll. Girhe, B.E., from September 2000 to September 2001, Kolhapur; 1♂, 3♀, Kolhapur 29-IX -2000, 2♂, 4♀, Malkapur 2-X-2000, 3♂, 5♀, Kolhapur 15-XI-2000, 1♂, 2♀, Kagal 17-XII-2000, 1♂, 5♀, Jaysingpur 2-II-2001, 1♂, 3♀, Gadhinglaj, 4-III-2001, 2♂, 4♀, Ajra, 5-IV-2001, 2♂, 5♀, Hatkanangale, 4-V-2001, 1♂, 4♀, Malkapur, 18-VIII-2001, 2♂, 5♀, Hatkanangale, 3-IX-2001; same data as above.

AEDES (STEGOMYIA) KOLHAPURENSIS SP. NOV.

Female (Fig. 35)

3.7 mm long, 0.65 mm broad, antenna 1.25 mm long, fore wing 2.10 mm long, hind leg 4.60 mm long; head dark brown, thorax blackish brown, abdomen slender and black.

Head

0.4 mm long, 0.35 mm broad, in dorsal view nearly rounded; vertex hairy, dark brown, vertical temporal bristles present; scales on vertex broad and flat; eyes rounded, black; interocular space 0.06

mm; proboscis 1.9 mm long, entirely dark, straight, cylindrical, maxillary palp 0.4 mm long; palpifer, palpus white; silver coloured scales or patches on clypeus and palpi; mandibles and maxillae straight, elongate and yellow with blade like stylet; nape blackish brown, rounded; antenna 1.25 mm long, blackish brown, 15 segmented, hairy and pilose; scape 0.12 mm long, 0.11 mm broad, black; pedicel 0.15 mm long, 0.10 mm broad; brown white, rounded; flagellum 0.98 mm long, 13 segmented.

Flagellar formula

1 L/W = 0.15, 13 L/W = 0.59, L 1/13 = 0.02, W 1/13 = 0.08.

Thorax

0.8 mm long, 0.65 mm broad, hump shaped scutum covers dorsal surface of thorax, undifferentiated laterally compressed and ornamented with few white scales; scutellum blackish brown, trilobed with three bunches of hairs on the lobes; scales on scutellum broad and flat; pronotum lack setae, bristles on mesonotum usually well developed; post spiracular bristles present; lyre shaped marking of white scales absent on thorax; prescutellar space blackish brown with prescutellar bristles.

Fore wing

2.10 mm long and 0.8 mm broad, wing unspotted only with dark scales, scales 0.02 mm long on veins on hind margin; squama semicircular, transparent with fringe of scales on margin; subcosta (SC) straight, 1.70 mm long reaching costa, R_{2+3} forked, R_{4+5} simple, no cross-vein connection of R_1 and R_2, Media (M) straight, bifurcated, 1.85 mm long, cross-vein r-m and m-cu both present, radius 2.0 mm long, straight, slightly curved at apex; cubitus (cu) bifurcated, cu_1 1.32 mm long, anal (A) vein 1.11 mm long, reaching margin.

Angle between m-cu and cu_1 94°, m_{3+4} and r-m 68° and R_{2+3} and r-m 33°.

Halter

0.18 mm long, 0.13 mm broad, rounded at tip.

Hind Leg

4.60 mm long, slender, banded and longer than body; coxa 0.6 mm long, broad and dark brown; trochanter 0.2 mm long, hard, black;.femur 1.10 mm long, cylindrical, speckled, tibia 1.3 mm long, dark and blackish brown; tarsus 1.40 mm long, five sigmented, four basal white bands on tarsal segment; claws small curved and toothed; pulvilli and empodium short.

Other legs

Special marks: Similar.

Abdomen

2.50 mm long, 0.60 mm broad, slender, black with dorsal transverse white and black bands, black and white patches irregularly distributed on turgal plates; abdomen densely hairy, anal cerci 0.27 mm long, 0.18 mm broad and hairy.

Colour

Black	:	Eye, scape.
Blackish brown	:	Legs, scutellum.
Brown	:	Vertex, pedicel.
White	:	Bands on legs, knee spot.
Yellow	:	Mandibles and maxillae.

Male

3.6 mm long, smaller than female, antenna plumose, brushy, phytophagous.

Host

Human.

Host plant

Unknown.

Holotype

Female, India, Maharashtra, Kolhapur, coll. Girhe, B.E., September 2000 to February 2001. Mouth parts, head, wing, legs and abdomen, mounted on slide labelled as above.

Paratype

6 male, 14 female; 1♂, 6♀, coll. Girhe, B.E., Ajra, 02-XI-2000; 3♂, 5♀, Hatkanangale, 19-XI-2000, Sex ratio (♂ : ♀) 1 : 2.3. Same data as above.

Discussion

Aedes kolhapurensis sp.nov. resembles with *Aedes aegypti* in Barraud's key in its characters.

1. Presence of fringe of hair on squama of wing.
2. Scales on vertex flat and broad.
3. Legs banded.
4. Claws curved and toothed.
5. Anal vein ends near VIth vein.

However, it differ from the above species by following characters:

1. Clypeus triangular
2. Vertex raddish brown with bristles.
3. Flagellar formula

   1 L/W = 0.15, 13 L/W = 0.59, L 1/13 = 0.02, W 1/13 = 0.08.
4. Anal cerci 0.27 mm long, rounded and blackish brown at tip.
5. Lyre shaped marking of white scales absent on thorax.

SUBGENUS *AEDES* MEIGEN 1818

Syst. Bescher. i.p. 13 Genotype, *A. cinereus* Mg.

Verrallina Theobald, 1903 M.C. iii p. 295.

Genotype, *Aedes butleri* Theo.

Neomacleaya Theobald, 1907, M.C. iv p. 238.

Genotype *N. indica* Theo.

Aioretomyia Leicestor, 1908, *Cul. malaya* p. 185.

Genotype, *A. varietas* Leic.

The subgenus *Aedes* erected by Meigen in 1818 is distinguished from other subgenera by having following characters:

1. Size small to medium.
2. Dark brown or reddish brown.
3. Mostly with lateral pale markings on abdomen.
4. Head entirely or mainly with flat scales.
5. Palpi very short in both sexes.
6. Scutellar scales always narrow.
7. Claw of fore and mid tarsi toothed in females and

8. Proboscis about length of fore femur or slightly longer.

Under this subgenus 18 species have been described from India (Barraud, 1934).

AEDES (AEDES) UNIFORMIS EDWARDS

Female (Fig. 36)

6.9 mm long, 1.40 mm broad, medium sized, antenna 1.80 mm long, black; fore wing 3.60 mm long, hind leg 9.55 mm long, head blackish brown, thorax brown, abdomen black.

Head

0.8 mm long, 0.55 mm broad, black brown; in dorsal view nearly rounded or globular, hairy, highly ornamented with spots and lines of snow white scales; eyes large, rounded and black; interocular space 0.10 mm; clypeus convex and ornamented with white scales; nape blackish brown, rounded; without any upright scales, tempus white; proboscis 2.3 mm long, about the length of fore femur, blackish brown, entirely dark; palpi 0.5 mm long, dark blackish brown, apical end of IVth palpus white, *Y4* length of proboscis; mandibles and maxillae straight, elongate, yellow, as long as proboscis and with apical blade like stylet; antenna 2.80 mm long, 15 segmented, blackish brown, pilose, hairy; scape 0.05 mm long, 0.13 mm broad; pedicel 0.15 mm long, 0.12 mm broad; circular and black; flagellum 2.60 mm long, 13 segmented.

Flagellar formula

1 L/W = 0.1,13 L/W = 0.38, L 1/13 = 0.03, W = 0.015.

Thorax

1.85 mm long, 1.20 mm broad, large, laterally compressed, undifferentiated, hump back appearance, ornamented with patches or spois of white scales; scutum covers dorsal region, black, with irregular white patches or spots; pronotum lack setae, bristles of mesonotum well developed, deep brown; 4 to 5 ppm bristles; mesothoracic spiracle rounded with spiracular bristles; sternopleuron and mesepimeron rectangular, flat and ornamented with irregular patches of white scales; post spiracular bristles present; antero and posterodorsocentral bristle present; scutellum trilobed with three bunch of hairs on the lobe, scales on scutellum white and flat.

Fore wing

3.60 mm long, 0.95 mm broad, unspotted with dark scales; scales 0.02 mm long, fringe of hairs on squama; small, dark scales on veins as well as fringe on hind margin; subcosta (SC) 3.12 mm long, straight, reaching the costa, R_{2+3} forked, R_{4+5} simple, no cross-vein connection of R_1 and R_2, Media (M) straight and bifurcated, marginally, 3.42 mm long, radius 3.56 mm long, slightly curved at tip, cross-vein r-m and m-cu very prominent, cubitus (cu) bifurcated, cu_1 1.76 mm long; anal (A) vein 1.46 mm long.

Angle between m-cu and cu_1 91°, m_{3+4} and r-m 83° and R_{2+3} and r-m 30°.

Halter

0.3 mm long, 0.2 mm broad, black and clubbed.

Hind leg

9.55 mm long, slender, banded and longer than body; coxa 1.0 mm long, blackish brown; trochanter 0.3 mm long, black and hard; femur 2.2 mm long, cylindrical with preapical white and apical end black; tibia 2.55 mm long, slender, clothed with dark scales; tarsus 3.5 mm long, five segmented, banded, first segment longer than others; claws curved and toothed; pulvillus short and hairy.

Other legs

Special mark: Tarsal claws of all legs toothed.

Abdomen

4.25 mm long, black, without pale marking except for small lateral patches of pale scales on tergite one; abdomen tapering, pointed and distinctly banded; anterior abdominal bands broad while last few narrowed in the middle, their hind margin 'A' shaped; a faint median longitudinal pale stripe absent; laterotergites blackish brown with few white scales, sternum yellow and with broad bands; anal cerci 0.4 mm long, 0.3 mm broad and hairy.

Colour

Black	:	Eye, abdomen
Blackish brown	:	Laterotergites, antenna
Brown	:	Wing
White	:	Femur
Yellow	:	Mandibles, maxillae
Yellowish brown	:	Thorax

Male

6.5 mm long, slender, shorter than female, antenna plumose and brushy, last two segments elongate, phytophagus.

Host

Human.

Host plant

Not known.

Holotype

Female, India, Maharashtra, Kolhapur, coll. Girhe, B.E., 12-XI-2000, Mouth parts, antenna, wing, legs, abdomen on slide, labelled as above.

Paratype

3 Male, 11 Female; Sex ratio (♂ : ♀), 1 : 3.6; coll. 1♂, 5♀, Sathe, T.V., Kolhapur, 20-XI-2000; 2♂, 6♀, Girhe B.E., 17-I-2001, Malkapur, same data as above.

AEDES (AEDES) INDICUS THEOBALD

Female (Fig. 37)

7.70 mm long, 1.30 mm broad, antenna 2.8 mm long, fore wing 4.6 mm long, hind leg 7.10 mm long; head blackish brown, thorax yellowish brown, abdomen brownish black.

Head

0.90 mm long, 1.10 mm broad, no upright scales on head, in dorsal view nearly rounded or globular, dark blackish brown; eyes rounded, large and black; interocular space 0.11 mm; clypeus smooth, convex, blackish brown; nape reddish brown, rounded; vertex semicircular, dark brown, tempus circular and brown; vertical and temporal bristles well developed; proboscis 2.4 mm long, dark brown, clothed with brown scales, labium elongated cylindrical while labium short and yellowish brown; palpi 0.6 mm long, four segmented, blackish brown and shorter than proboscis; mandible and maxillae elongated, slender with blade like stylet, as long as proboscis; antenna 2.8 mm long, blackish brown, 15 segmented, hairy, pilose; scape 0.22 mm long, 0.17 mm broad, black; pedicel 0.30 mm long, 0.13 mm broad, brown and rounded; flagellum 2.28 mm long, 13 segmented.

Flagellar formula

1 L/W = 0.16, 13 L/W = 0.062, L 1/13 = 0.029, W 1/13 = 0.012.

Thorax

2.10 mm long, 1.30 mm broad, undifferentiated, laterally compressed, yellowish brown; scutum yellowish brown, covers dorsal area, hump like appearance; mesothoracic and metathoracic spiracle rounded, with spiracular bristles; anterior and posterior pronotallobes semi circular, dark brown; scutellum yellow, trilobed, with three bunch of hairs on the lobe; bristles on mesonotum well developed, pronotum lack setae; sternopleuron flat, rectangular, larger than mesepimeron; metepisternum, metepimeron and metameron forms metapleuron; thorax with dark brown stripe on each side.

Fore wing

3.6 mm long, 1.3 mm broad, unspotted with only dark scales, squama semicircular; transparent with fringe of hairs, dark scales present; scales 0.24 mm long, present on veins, fringe on hind margin; subcosta (SC) vein 3.80 mm long, straight; R_{2+3} forked, R_{4+5} simple, no cross-vein connection of R_1 and R_2, Media (M) 3.95 mm long; radius 4.50 mm long, cross vein r-m and m-cu both present; cubitus (cu) bifurcated, cu_1 2.08 mm long, anal (A) vein 1.59 mm long.

Angle between m-cu and cu_1 93°, m_{3+4} and r-m 75° and r_{2+3} and r-m 35°.

Halter

0.25 mm long, 0.22 mm broad, brown and flat at tip.

Hind leg

7.10 mm long, slender and unbranched, blackish brown, coxa 0.9 mm long, broad, hairy, yellow; trochanter 0.20 mm long, broad, hard; femur 1.40 mm, cylindrical with dark scales; tibia 2.60 mm long, slender, with tibial dirty yellow scales and yellow tibial spines; tarsus 3.00 mm long, five segmented, first tarsus longer than other and unbanded; claws curved, toothed; pulvilli and empodium short, hairy.

Other legs

Special marks: Similar.

Abdomen

4.70 mm long, 1.3 mm broad, long tapering and pointed;

brownish black, marked with complete transverse bands on tergites 2nd and 4th; abdominal dorsal transverse white bands broad and constricted mid dorsally; anal cerci 0.3 mm long and 0.2 mm broad, brown; in dorsal view, tergum of abdominal segment sparsely covered with flat scales.

Colour

Black	:	Eye.
Blackish brown	:	Clypeus, scutum.
Brown	:	Vertex
Yellow	:	Tibial spines, mandibles, maxillae. Scutellum.
Yellowish brown	:	Coxa.

Male

7.40 mm long, slender, smaller than female, antenna plumose, brushy with last two segments elongate.

Host

Human.

Host plant

Grass and other vegetation.

Holotype

Female, India, Maharashtra, Kolhapur coll. Girhe, B.E., Kolhapur. 29-IX-2000, mouth parts, head, wing, legs and abdomen on slide, labelled as above.

Paratype

10 male, 36 female, Sex ratio (♂ : ♀) 1 : 3.6. coll. Girhe, B.E., September 2000 to September 2001, Kolhapur; 4♂, 15♀, Kolhapur, 20-X-2001; 6♂, 21♀, Panhala, 21-V-2001; same data as above.

AEDES (AEDES) SANGITI SP. NOV.

Female (Fig. 38)

7.60 mm long and 1.05 mm broad, antenna 2.60 mm long, fore wing 4.65 mm long, hind leg 9.25 mm long, abdomen 4.69 mm long, head yellowish brown, thorax brown, abdomen brown.

Head

0.31 mm long, 0.40 mm broad, in dorsal view globular, dorsal surface of head with many flat scales; clypeus rounded whitish; vertex smooth with white margin; interocular space 0.25 mm; proboscis 3.5 mm long, dark unspeckled; labellum brown; maxillary palp 1.23 mm long and four segmented; palpus short and broad, apical palpus longer than others; mandible and maxillae long, blade like, stylet round and curved; antenna 2.60 mm long pilose, 15 segmented, scape 0.12 mm long, 0.15 mm broad; pedicel 0.07 mm long, 0.12 mm broad, yellowish brown; flagellum 2.41 mm long, 13 segmented.

Flagellar formula

1 L/W = 0.03, 13 L/W = 0.50, W 1/13 = 0.14, L 1/13 = 00.04.

Thorax

1.60 mm long, 1.10 mm broad undifferentiated, blackish brown, laterally compressed, narrow with hump backed appearance, with whitish pale hairs; thoracic greater part of the dorsal surface formed by shield like scutum of the second segment which bears the wing; scutellar scales broad and flat; anterior laterval pronotallobe leaf like, whitish, posterior notal lobe rounded with few white scales, scutum with verticle scutal suture, scutellum trilobed with white and narrow scales and three bunches of hairs on the lobes; prealar knob black, mesoepimeron and mesomeron blackish brown with sparsely white and black scales; metapleuron yellowish brown, metepisternum triangular and brownish, metepimeron verticle slit like; metameron triangular , lies at the base of metaepimeron.

Fore wing

4.65 mm long, 2.10 mm broad, elongated, unspotted, bears tiny dark scales on veins on a fringe on hind margin; subcosta straight, 4.20 mm long, Media (M) bifurcated, 3.60 mm long; radius straight, slightly curved at apex, 4.35 mm long; cubitus bifurcated, cu_1 1.89 mm long, anal vein 1.63 mm long.

Angle between m-cu and cu_1 95°, m_{3+4} and r-m 85° and R_{2+3} and r-m 39°.

Halter

0.30 mm long, 0.21 mm broad, brown and rounded at tip.

Hind leg

9.25 mm long, unbanded, longer than body; coxa 0.60 mm long, broad, yellowish in colour; trochanter 0.12 mm long, triangular and hard; femur 2.21 mm long, elongated, cylindrical, blackish, brown dorsally and whitish ventrally with femoral spurs, white knee spot at apical end; tibia 3.10 mm long, elongated straight, slender unspeckled, black brown scales and brownish tibial spines silvery; tarsus 3.22 mm long, 5 segmented, unbanded, elongated; tarsal claws curved, toothed; pulvillus and empodium narrow.

Other legs

Special marks: Similar.

Abdomen

4.69 mm long, 1.10 mm broad, tapering and pointed towards posterior, with white and brownish scales, dorsal surface not distinctly banded, pale brown band lies ventrally, blackish brown scales dorsally; first seven segments dark brown dorsally and yellowish brown ventrally; VIIIth abdominal segment narrow, non retractile, completely white with fine hairs; anal cerci 0.11 mm long, 0.08 mm broad, rounded and black at tip.

Colour

Black	:	Dorsum of abdomen.
Brownish black	:	Thorax, leg, palpi, wing.
Brown	:	Labellum, head.
White	:	Tip of hind femur, VIII segment of abdomen.
Yellow	:	Maxillae, mandible.
Yellowish brown	:	Ventral stripes of abdomen.

Male

7.4 mm long, smaller than female, phytophagous.

Host

Man, cow, buffalow, grass and other vegetation.

Holotype

Female, India, Maharashtra, Kolhapur coll. Girhe B.E., 15-X-2000, head, legs; wing, abdomen mounted on slide, labelled as above.

Paratype

46 female, 10 male, Sex ratio (♂ : ♀), 1 : 4.6, from September, 2000 to September 2001, Kolhapur; 11♂, 2♀, Coll. Girhe, B.E., Hatkanangale; 5♂, 3♀, coll. T.V. Sathe, Jaysingpur, 21-XI-2000, same date as above.

Discussion

Aedes sangiti sp. nov. resembles with *Aedes stegomyia* in Barraud's key in its characters.

1. Abdominal segment VIII not completely retractile,
2. Tarsal claws toothed in fore leg and mid leg,
3. Dorsal surface of head with many flat scales,
4. Scutellar scales broad and flat,
5. Proboscis dark.

However, it differs from the above species by following characters:

1. Vertex smooth, with white border.
2. Hind tibia with silvery spines.
3. Flagellar formula

   1 L/W = 0.03, 13 L/W = 0.50, W 1/13 = 0.04, L 1/13 = 0.14.
4. Abdomen segments (1 to VII) not distinctly banded and covered densely with blackish brown scales.
5. VIIIth abdominal segment completely white.
6. Anal cerci 0.11 mm long, 0,07 mm broad, rounded and black.

AEDES (AEDES) PUNCHGANGI SP. NOV.

Female (Fig. 39)

7.35 mm long and 0.10 mm broad, antenna 2.80 mm long, fore wing 2. 1 mm long, hind leg 10.5 mm long; head yellowish brown, thorax brown, abdomen pale yellowish white.

Head

0.5 mm long, 0.45 mm broad; in dorsal view globular with few flat scales; eyes large, black; interocular space 0.21 mm, clypeus rounded, brownish with marginal bristles; vertex smooth and yellowish brown; nape yellowish brown and triangular; maxillary palpi very short; proboscis 2.00 mm long, brown, straight, about length of fore femur; labium elongated and brownish; labellum yellowish brown, mandibles and maxillae long, blade like stylet;

antenna 2.80 mm long, blackish brown, pilose, hairy, 15 segmented; scape 0.10 mm long, 0.11 mm broad; pedicel 0.06 mm long, 0.08 mm broad, rounded and brown; flagellum 2.64 mm long, 13 segmented.

Flagellar formula

1 L/W = 0.12,13 L/W = 0.60, L 1/13 = 0.02, W 1/13 = 0.12.

Thorax

2.35 mm long, 1.80 mm broad, narrow, laterally compressed brown with stripe on each side, scutum shield like, covers dorsal surface; scutellum yellowish and triangular; acrostichal bristles elongated and lies in the mid dorsal surface of scutum; prescutellar space pale yellow with prescutellar bristles; mesothoracic and metathoracic spiracles with spiracular bristles; postnotum pale yellow and cylindrical; propleuron and sternopleuron rounded and dark brown.

Fore wing

4.15 mm long, 1.00 mm broad, unspotted bears tiny dark scales along the veins as well as fringe of them along the hind margin, fringe scales leaf like; subcosta (SC) 3.88 mm long, straight, reaching the costa; base of subcosta without tufts of hair on the underside of wing; media (M) bifurcated, 3.98 mm long, cu_1 2.87 mm long, radius 4.13 mm long; anal (A) 1.62 mm long, reaching margin.

Angle between m-cu and cu_1 96°, m_{3+4} and r-m 81° and R_{2+3} and r-m 35°.

Halter

0.21 mm long, 0.17 mm broad, yellow and expanded at tip.

Hind leg

10.50 mm long, unbanded, yellowish brown; coxa broad, 1.00 mm long; trochanter 0.3 mm long, hard and brown; femur 2.00 mm long, cylindrical, basal end whitish yellow while apical end yellowish brown; knee pale yellow; tibia 3.10 mm long, slender, brown with dark brownish scales; apical end of tibia with tibial spurs, no tarsal ring on tarsal segment; tarsus 4.10 mm long, unbanded, elongated, claw curved and toothed, empodium and pulvillus narrow.

Other legs

Similar.

Abdomen

4.50 mm long, 2.11 mm broad, cylindrical tapering at hind end and sparsely covered with over lapping scales; scales lies on transverse margin of the segments; first to VIth abdominal segments whitish while VIIth and VIIIth abdominal segments brownish and hairy; post genital plate brownish at tip, laterotergites yellowish with few scales; anal cerci 0.35 mm long, entirely brown, leaf like, hairy.

Colour

Black	:	Eye
Blackish brown	:	Vertex, antenna
Brown	:	Cerci, post genital plate and proboscis, Thorax
Yellowish brown	:	Labellum
Yellow	:	Laterotergites, coxa, prescutellar space, maxillae and mandibles.
White	:	First six segment of abdomen.

Male

7.1 0 mm long, smaller and slender than female, phytophagous.

Host

Man.

Host plant

Grass and other unknown vegetation.

Holotype

Female, India, Maharashtra, Kolhapur, Coll. Girhe, B.E., 30-XII-2000, head, antenna, wing, legs abdomen mounted on slide labelled as above.

Paratype

5 male, 25 female: Sex ratio (♂ : ♀) 1: 5; 2♂, 10♀ Coll. Girhe, B.E., September 2000 to September 2001; Kolhapur; 1♂, 6♀ Coll. Girhe, B.E., Panhala; 6-I-2001; 2♂, 9♀ Coll. Sathe T.V., Kolhapur, 21-XII-2000; same data as above.

Discussion

Aedes punchgangi sp.nov. resembles with *Aedes dorsalis* in Barrauds key in its characters:

1. Dorsal surface of head with flat scales.
2. Proboscis brown.
3. Thorax dark brown with stripe on each side.
4. No pale ring on tarsi.
5. Abdomen long tapering and pointed, claws toothed.

However, it differs from the above species by following characters:

1. Head, thorax brown while abdomen white.
2. Flagellar formula
   1 L/W = 0.12,13 L/W = 0.60, L 1/13 = 0.02, W 1/13 = 0.12.
3. First six (I to VI) abdominal segments white while VIIth and VIIIth segment brownish and hairy.
4. Anal cerci 0.35 mm long dark brown.

SUBGENUS *MUCIDUS* THEOBALD, 1901

M.C.i; p. 268. Genotype *Culex alternans* Westwd.

Pardomyia Theobald, 1907, M.C. iv, p. 280.

Genotype, *P. aurantia* Theo.

Elcrinomyia leicester, 1908, Cul. Malaya, p. 71.

Genotype, *E. aurcostriata* Leic.

The Subgenus *Mucidus* Theobald is distinguished from other subgenera by having following features:

1. Adults large sized with outstanding yellow, white and brown scales on body and legs, giving them mottled and mouldy appearance.
2. Wing mottled with yellow, brown and creamy scales.
3. Mesonotum remarkably long and twisted, resembling strands of cotton-wool.
4. ppn bristles numerous, usually about 20.

From India only two species have been described under this subgenus. From Kolhapur district a new species, *Aedes indica* sp. nov. have been reported and described for the first time.

AEDES (MUCIDUS) INDICA SP. NOV.

Female (Fig. 40)

Large mosquito, 10.30 mm long, fore wing 7.6 mm long, hind leg 19.5 mm long, antenna 5.2 mm long and yellowish dirty brown and longer than proboscis; head yellowish brown, thorax pale yellow, abdomen yellowish brown.

Head

1.10 mm long, 1.00 mm broad, in dorsal view globular, compound eyes black, vertex dark brown, white scales on anterior margin of vertex; clypeus blackish brown and rounded; nape black; proboscis brownish gray, 4.6 mm long, white scales present on the middle region; labellum dark brown; maxillary palp 2.3 mm long, not more than 1/2 length of proboscis, hairy, pale yellow; antenna 5.2 mm long, yellowish brown, pilose; scape black, 0.25 mm long, 0.30 mm broad; pedicel 0.30 mm long, 0.20 mm broad and brown; flagellum 4.65 mm long, 13 segmented.

Flagellar formula

1 L/W = 0.35, 13 L/W = 0.88, L 1/13 = 0.08, W 1/13 = 0.02.

Thorax

3.00 mm long, 2.00 mm broad, laterally compressed, hump shaped, brown, not with light matt, with brownish integument; scutellum trilobed with bunch of hairs on each lobe; mesonotum remarkably long and twisted, mesonotal scales whitish, long and twisted, mesonotal bristles light brown.

Fore wing

7.5 mm long, 4.2 mm broad, yellowish with dark spots, wing mottled with yellow, brown and with creamy scales, ppn bristles numerous; no large yellow stick towards apex of wing extending on vein 1. Subcosta straight, 6.5 mm long, reaching costa, costa yellow with dark scales on entire region except basal parts; radius 7.3 mm long, straight, slightly curved at apex, R_{2+3}, R_{4+5} simple, no cross-vein connection of R_1 and R_2, Media (M) straight and bifurcated apically, cross-vein r-m and m-cu prominent; cubitus (Cu) forked, cu_1 3.9 mm; anal (A) vein 3.2 mm long, approaching margin; hind margin of wing with dark and yellow scales placed alternately.

Angle between m-cu and cu_1 91°, m3+4 and r-m 88° and R_{2+3} and r- m43°.

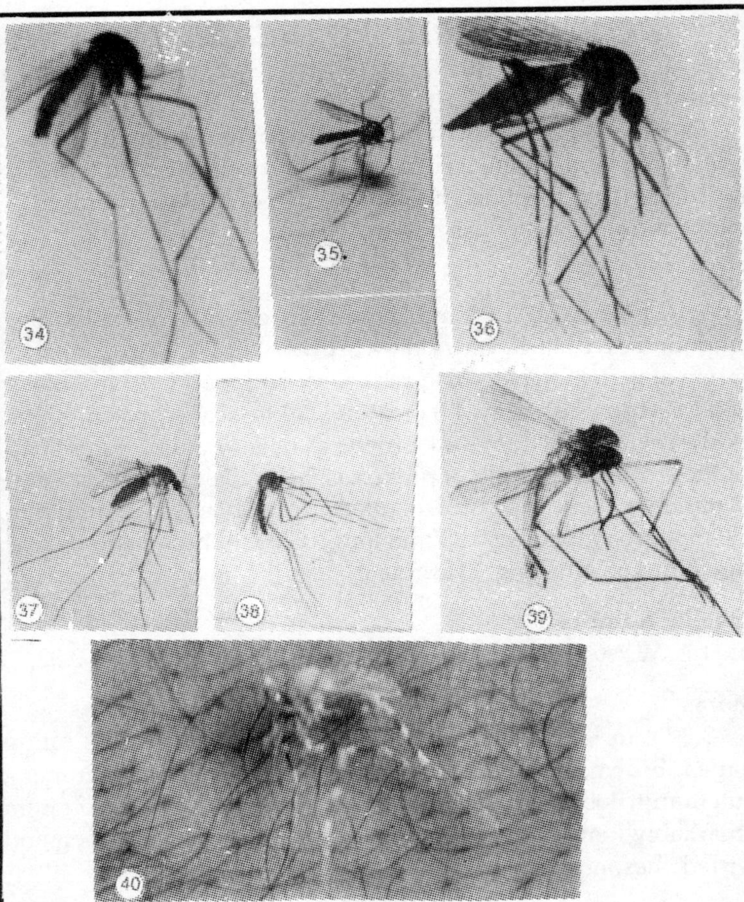

Plate V

Fig. 34: *Aedes aegypti*, Fig. 35: *Aedes kolhapurensis*, Fig. 36: *Aedes uniformis*, Fig. 37: *Aedes indicus* sp. nov., Fig. 38: *Aedes sangiti* sp. nov., Fig. 39: *Aedes punchgangi* sp. nov., Fig. 40: *Aedes indica* sp. nov.

Halter

0.45 mm long, 0.38 mm broad, brownish, clubbed.

Hind Leg

19.5 mm long, slender and longer than body; hind coxa 1.5 mm long, broad with white and yellow scales; hind trochanter 0.6 mm long, broad, black; hind femur 4.0 mm long with two yellow bands and brown bristles; tibia 5.00 mm long with three mid white rings and brownish tibial bristles; tarsus 8.4 mm long, four white bands present on tarsus.

Other legs

Special marks: all legs brown, banded with white stripes.

Abdomen

6.20 mm long, 2.5 mm broad, brown with white scales, predominating at middle; last two segments slender and elongated than other; abdomen banded; anal cerci 0.35 mm long, rounded and black.

Colour

Black	:	Eye.
Brown	:	Vertex, proboscis, labellum, thorax
White	:	Rings on legs. scales on abdomen and , thorax
Yellow	:	Maxillary palp, scales on wing, hind tarsus.
Yellowish Brown	:	Antenna.
Blackish Brown	:	Clypeus.

Male

9.8 mm long, slender, antenna plumose and brownish.

Host

Human.

Host plant

Unknown.

Holotype

Female, India, Maharashtra, Kolhapur. coll. T. V. Sathe, 26-VII-2001, Shivaji University, Kolhapur; head, antenna, wing hind leg and abdomen on slide, labelled as above.

Paratype

Sex ratio (♂ : ♀) 1 : 1; 1♂, 1♀ Coll. B. E. Girhe, 14-VIII-2001, same data as above.

Discussion

Aedes Indica sp. nov. resembles with *A. M. seatophagoides* Theobald by having large size and leg markings, proboscis with white scales at middle (Barrand, 1934). However it differs from above species by having following characters:

1. Pro boscis brownish gray.
2. Palpi not more than 1/2 length of proboscis.
3. Abdomen with white scales predominating at middle than tip.
4. Four white bands on tarsus.
5. Dark scales on entire costa but lacking at basal region.
6. Wing length and breadth 7.5 mm and 4.2 mm respectively.
7. Larger body size.

Chapter 5
Seasonal Abundance and Distribution

Introduction

Ecology plays an important role in pest management. The concept of ecological pest management is emerging fastly in pest control strategies. Hence, information on ecological aspects of a target species is essential. Geographical distribution, seasonal abundance, habitat, natural enemies, climatic factors are important aspects of ecology of a species. The population of a species is influenced by biotic and abiotic factors. Biotic factor refers to parasites, predators, pathogens, diseases, competitions etc. and abiotic factor refers to temperature, humidity, rainfall, light, wind, topography etc. The seasonal abundance has relation with biotic and abiotic factors. Hence this aspect has great importance in ecology and pest management.

Sampling of insect population is an important task in population dynamics which estimate the number of species present in a target area. The information on seasonal abundance of a species suggests the time of occurrence and probable number of the species available in a specific region. Further, it helps in adopting appropriate control measures of target species. The information on seasonal abundance also help in understanding useful and harmful role of mosquito species in human health care of a region. Keeping in view all above facts the present work was carried out.

Review of literature showed that Beklemisher and Shipitsina (1957), Bhatia *et. al.* (1958), Foote and Cook, (1959), Das *et. al.* (1984), Rao, (1984), Nagpal and Sharma (1987) etc. have contributed on seasonal abundance of mosquitoes in India. However, very little attention is paid on the seasonal abundance of mosquitoes in Maharashtra except scanty information of health care department.

Materials and Methods

The seasonal abundance of mosquitoes have been studied in different places of Kolhapur district namely, Hatkanangale, Kagal, Ajra, Gadhinglaj, Jaysingpur, Panhala and Malkapur. Mosquitoes were collected with the help of suction tube (Fig. 7, 9) for 15 minutes at 15 days interval. The specimen collected during study period were identified by consulting appropriate literature and preserved in Department of Zoology, Shivaji University, Kolhapur. Distributional record of the species have also been studied by visiting and collecting the specimen from above places of Kolhapur district and previous distributional record (Table 1) is given by consulting appropriate literature.

Results

The observations on seasonal abundance of mosquito species belonging to genera *Anopheles, Culex* and *Aedes* have been given in Table 1 and distributional record of the species is given in Table 2. The results indicate that the species *Anopheles culicifacies, Culex pipiens* and *Aedes aegypti* were abundant through out the year while *Aedes indica* sp.nov. was extremely rare. Other species recorded in Kolhapur district were moderate in their occurrence.

Discussion

Success of controlling mosquito borne diseases is dependant on rapidity with which a situation can be currently evaluated. This, inturn, depends on having or quickly obtaining an accurate population density of potential vector of a disease in the area under consideration. The rate of mosquito borne diseases depends on the index of species of the region. The rapid separation and identification of mosquitoes of primary medical importance is an important step in the assessment of disease potential of an area. Review of literature indicates that Foote and Cook (1959), Rao (1984), Nagpal and Sharma (1987, 1995) studied the seasonal abundance of mosquitoes of medical importance in India.

Foote and Cook (1959) designed 32 regions of mosquitoes in the World. India and Ceylon is one of the world region which they visualized. Under this region they studied mosquito borne diseases, namely malaria, yellow fever, dengue, encephalitis and filariasis. They reported that the most important and wide spread vector of malaria in Central and Peninsular India is *Anopheles culicifacies.* This species was also the most important vector in Srilanka. In the foot hill areas of Peninsular India, *Anopheles fluviatilis* species was dominant for causing hyperendemic malaria. Foote and Cook (1959), also reported *Anopheles varuna* from East-central India, *Anopheles stephensi* from Northern West-coast and Ganges plain, *Anopheles sundaicus* from Calcutta, southward to the middle of Peninsular India along the East coast and *Anopheles annularis* from Orissa.

Nagpal and Sharma (1987, 1995) reported 14 species of the genus *Anopheles* from different parts of India while, Rao (1984) reported 51 species of *Anopheles* and their subspecies and varieties from India giving type locality, taxonomy, distribution, adult bionomics, larval ecology and diseases. However, from Maharashtra State there is no published comprehensive list of *Anopheles* species. In the present study from Kolhapur district, Maharashtra four species namely, *Anopheles culicifacies, An. stephensi, An. theobaldi* and *An. subpictus* have been reported. The most prevalent species of *Anopheles* in Kolhapur region was *Anopheles culicifacies* and *Anopheles subpictus* was rare in the region.

As like other states of India, malaria is common disease in Maharashtra, its incidence was studied during the years, 1992-2000 by Girhe and Sathe (2001). They reported that the incidence of malaria during the year 1992-1996 was increasing in order. Maximum, 700 infection cases have been reported during the year, 1996. However, the incidence of malaria declined from the years 1997-2000 by showing only 60 infection cases during the year 2000. The reasons for decline in malaria incidence were the proper strategies of mosquito control adopted by the health department and low rainfall in the region.

Foote and Cook (1959) mentioned the following centres of filariasis in India, the lower parts of the Ganges River basin, some sea-coast areas in Bihar and Orissa on the northeast coast and Travancore state in the south indicating prevalence of the *Culex annulifera,* was the chief vector. In the present study, four species of *Culex* namely *Culex epidesmus, Culex pipiens, Culex modestus, Culex malayi* have been reported. As far as filariasis infection through mosquitoes is concern no report is available from Kolhapur region.

Table 1: Distribution Record of Mosquitoes

Sr. No.	Species	Distribution record in Kolhapur District	Previous Distribution record in India
(A) Genus *Anapheles*			
1.	*An. Culcifacies*	Kolhapur, Malkapur, Jaysingpur, Hatkanangale, Ajra, Panhala, Gadhinglaj, Kagal	West Bangal, Arunachal Pradesh, Utter Pradesh, Assam, Orissa, Tamilnadu, Maharashtra etc.
2.	*An. stephensi*	Kolhapur, Malkapur	Punjab, Uttar Pradesh, Bihar, Delhi, Goa, Gujarat, Haryana, Kashmir etc.
3.	*An. theobaldi*	Kolhapur, Malkapur, Gadhinglaj	Orissa, West Bangal, Uttar Pradesh, Madras, Maharashtra etc.
4.	*An. subpictus*	Kolhapur, Malkapur	Andaman Islands and Lakshadweep, Maharashtra etc.
(B) *Genus Culex*			
1.	*Culex epidesmus*	Kolhapur, Hatkanangale, Gadhinglaj	Not known
2.	C. *pipiens*	Kolhapur, Kagal, Hatkanangale, Malkapur, Panhala, Ajra, Gadhinglaj, Jaysingpur	West Bangal, Uttar Pradesh,
3.	C. *modestus*	Kolhapur, Hatkanangale	Not Known
4.	C. *malayi*	Kolhapur, Kagal, Ajra	Not Known
(C) Genus *Aedes*			
1.	*Aedes aegypti*	Kolhapur, Kagal, Ajra, Panhala, Hatkanangale, Jaysingpur, Malkapur	Delhi, Madras, Assam etc.
2.	*A. kolhapurensis* sp. nov.	Kolhapur, Hatkanangale	Not Known
3.	*A. uniformis*	Kolhapur, Panhala	Not known
4.	*A. indicus*	Kolhapur, Panhala	Not Known
5.	*A. sangiti* sp.nov.	Kolhapur, Hatkanangale	Not Known
6.	*A. punchgangi* sp. Nov.	Kolhapur, Panhala	Not Known
	A. India sp.nov.	Kolhapur	Not Known

Table 2: Seasonal, Abundance of Mosquitoes

Sr. No.	Mosquito Species	Jan.	Feb.	March	April	May	June	July	Aug.	Sept.	Oct.	Nov.	Dec.
1.	*Anopheles culicifacies*	+	+	+	+	+	+	+	+	+	+	+	+
2.	*Anopheles stephensi*	–	–	–	–	–	–	+	+	+	–	–	–
3.	*Anopheles theobaldi*	–	–	–	–	–	+	+	–	+	–	–	–
4.	*Anopheles subpictus*	–	–	–	–	–	–	–	+	+	–	+	–
5.	*Culex epidesmus*	+	–	+	–	–	–	–	+	–	–	–	–
6.	*Culex pipiens*	+	+	+	+	+	–	+	+	+	+	+	+
7.	*Culex modestus*	–	–	–	–	–	+	–	–	+	–	–	–
8.	*Culex malayi*	+	+	+	–	–	–	–	–	–	–	–	–
9.	*Aedes aegypti*	–	+	+	+	+	–	–	+	+	+	+	+
10.	*Aedes kolhapurensis* sp.nov.	–	+	–	–	–	–	–	+	–	–	–	+
11.	*Aedes uniformis*	+	–	–	–	–	–	–	–	–	–	–	–
12.	*Aedes indicus* sp. nov.	–	–	–	–	+	–	–	–	+	+	–	–
13.	*Aedes sangiti* sp. nov.	–	–	–	–	–	–	–	–	–	–	+	+
14.	*Aedes punchgangi* sp. nov.	–	–	–	–	–	–	–	–	–	–	+	+
15.	*Aedes indica* sp. nov.	–	–	–	–	–	–	+	+	–	–	–	–

Foote and Cook (1959) reported that *Aedes aegypti* was common through out India. It was especially common after the rainy season in West Bengal and Madras but, it was less so in Western India and practically unknown in the mountain provinces along the northern western frontier. The other species, *Aedes albopictus* reported by them was not so closer to man and was less universally distributed in India. From Maharashtra still a comprehensive list of *Aedes* species is to be published. The present work includes the first record of seven species of genus *Aedes* from Kolhapur district of which three species, *Aedes aegypti, A. indicus* and *A. uniformis* are known to science. However, *Aedes kolhapurensis* sp. nov. , *A. indica* sp. nov., *A. punchgangi* sp. nov. and *A. sangiti* sp. nov. are newly described and reported from India.

Chapter 6
Conclusion

Human health is affected by a variety of organisms which include insects and other pathogens. The insects related to epidemiology include mosquitoes, house flies, sand flies, rat flea, tse-tse fly etc. Among these insects, mosquitoes play an important role in epidemiology as they cause fetal diseases like malaria, filariasis, encephalitis, yellow fever, dengue etc.

Mosquitoes are diversified taxonomical group of insects. Biodiversity of mosquitoes is one of the important aspects of medical sciences which is directly related to human diseases. Hence, in the present study, biodiversity with respect to the taxonomy, seasonal abundance and distribution have been studied from Kolhapur, district of Maharashtra.

Kolhapur district comes under Western Maharashtra and is characterized by several kinds of water bodies, hilly regions and Western Ghats. The river Panchganga, Panchganga river system and heavy rainfall favours mosquitoes for breeding in the region which ultimately cause diseases like malaria, dengue etc. to human beings.

The book have been divided into six chapters. The first chapter is devoted to the introduction which provides national and international status of the topic and review of literature. The second chapter deals with materials and methods. The third chapter is devoted for mosquito borne diseases. The fourth chapter narrates the biodiversity of mosquitoes with respect to taxonomy. In all 15 species belonging to the genera *Anopheles, Culex* and *Aedes* have been studied under this chapter. Four species of the genus *Aedes* have been described for the first time from India. The fifth chapter embodies the seasonal abundance and distribution of 15 species in this region. The results indicate that *Anopheles culicifacies, Culex pipiens* and *Aedes aegypti* were abundant throughout the year while *Aedes indica* sp. nov. was rarely found and reported for the first time. The sixth chapter embodies the conclusion of the present work and the seventh chapter refers to bibliography.

Bibliography

Afridi, M.K. and Puri, I.M. (1940) Studies on the behaviour of adult *Anopheles culicifacies. J. Mal. Inst. Ind.*, 3 : 1-22.

Atwal, A.S. (1933) *Agricultural pests of India and South east Asia*. Kalyani Publ., 1-289 pp.

Ananthakrishnan, T.N. (1993) Biological diversity concepts and conservation strategies. *Hexapoda,* (2) : 5-8.

Apiwathnasom Chamnarn. (1986) A list of mosquito species in Southeast Asia. Museum and Reference Centre, I.S.B.N., 974-585-965-6, : 1-73 pp.

Barraud, P.J. (1934) The Fauna of British India including Ceylon and Burma, V, Taylor and Francis, London. 1-463 pp.

Basu, P.C. (1958) A note on malaria and filariasis in Andaman and Nicobar Island, *Bull. Nat. Soc. Ind. Med. Mosq. Dis., 6*: 193-206.

Bates, Marston, (1949) The Natural History of Mosquitoes. New York: The Macmillan Company XV, 378.

Beklemisher, W.N. and Shipitsina, N.K., Eds. (1957) Seasonal phenomena in the life of the anophelines of the U.S.S.R. (Institute of Medical Parasitology, Medical Parasitology, Moscow, U.S.S.R., 518.

Bhatia, M.L., B.L. Wattal, M.L Mannen and N.L. Karla (1958) Seasonal prevalence of anophelines near Delhi, *Ind. J. Mal.*, 12 : 13-38.

Busvine, J.R. (1980) Insect and hygiene: The biology and control of insect pests of medical and domestic importance, 521-527 pp.

Carpenter, S. J., and Lacasse, Walter J. (1955) *Mosquitoes of Northern America* (North of Mexico). Berkely : Univ. of California Press VII, 360 pp.

Cheng, T.C. (1964) The biology of animal parasite W.B.S. Camp., London and Top, Comp. Japan, 589-602 pp.

Christophers, S.R. (1933) The Fauna of British India including Ceylon and Burma, Diptera Vol. VI, Family Culicidae, Tribe Anopheline, Taylor and Francis, London, 1-371 pp.

Christophers, S.R. (1960) *Aedes aegypti* (L.) the Yellow Fever Mosquito, Cambridge Univ. Press, Cambridge, Engl; 739 pp.

Das, S.C., Das, N.G. and Baruah, I. (1984) Mosquito Survey in Meghalaya, *Indian J. Publ. Hlth.*, (3) : 147-151.

Das, B.P., Rajagopal, R. and Akiyama, J. (1990) Pictorial key to the species of Indian anopheline mosquitoes, *Zoology*, 2(3) : 131-162.

Edwards, F.W. (1932) Diptera, Family Culicidae, Brussels. Genera Insectorum, de P. Wytsman, 194 me Fascicule. 258 pp.

Evans, A.M. (1938) Mosquitoes of the Ethiopian region II, Anophelini adults and early stages, British Museum of Natural History, London, 404 pp.

Foote, Richard H. and Cook, David R. (1959) Mosquitoes of Medical Importance, U.S. Dept. Agric., Agric. Handbook No.152, 158 pp.

Girhe, B.E. and Sathe, T .V. (2001) Incidence of Malaria in Kolhapur district, Maharashtra, *Nat. Sym. Devt. Environ. and Human Conds.* No. 3.6, 36pp.

Gordon, R.M. (1992) Entomology of Medicine, Blackwell Scientific Publication, Oxford, 1-337pp.

Harrison, B.A., Scanlon, J.E. and Reid, J.A. (1973) A new synonym and new species name in the Southest Asian *Anopheles hyracanu*, Complex. *Mosq. Systematics*, 5: 263-268 pp.

Horsfall, William R. (1955) Mosquitoes, their bionomics and relation to diseases, New York : The Ronald Press Company, 723 pp.

James, S.P. and Liston, W.G. (1911) *A monograph of Anopheline mosquitoes of India*, Thacker Spink and Co., Culcutta, 127 pp.

Kain, W .D. (2000) Population Education for Population stabilization, Family Planning Association of India, Vol. 29 : 16 pp.

Knight, K.L. and Stone, A. (1977) A catalogue of the mosquitoes of the World (Diptera : Culicidae), *The Geo. W. Kng. Comm.. Baltimore. Maryland*, 21202, 611 pp.

Knowles, R. and White, R. (1927) *Malaria, its investigation and control*, Thacker Spink and Co., Calcutta.

Korke, V.T. (1928) Observations on Filariasis in some area in British India I and II, *Indian Jour. Med. Res.*, 16: 717-732.

Korke, V.T. (1932) Observations on Filariasis in some areas in British India, VIII, *Indian Jour. Med. Res.*, 20 : 335-339.

Mattingly, P.F. (1962) Mosquito behaviour in disease eradication, *Ann. Rev. Entomol.*, *7*: 419-436.

Mitzmain, Robert, (1944) *Handbook of Mosquitoes of North America*, 2nd ed. Ithaca : Comstock Publ. Co. Inc. VIII, 314 pp.

Mitzmain, M.B. (1917) "Anopheline mosquitoes, their distribution and infection under field conditions", U.S. Public Health Service, *Pub. Health Rep.*, 32 : 536-40.

Nair, K.S.S. and Mathew, G. (1993) Diversity of insects in Indian forests: The state of our knowledge, *Hexapoda*, 5(2): 71-78.

Nagpal, B.N. and Sharma, V.P. (1987) Survey of Mosquito fauna of Northeastern region of India, *Indian J. Malariol.*, *1*: 143-149.

Nagpal, B.N. and Sharma, V.P. (1995) Indian *Anopheles*, Oxford and I.B.H. Publishing Co. Pvt. Ltd., 1-416 pp.

Puri, I.M. (1947) The practical application of D.D.T. for Malaria control in rural and urban areas in India, *Ind. J. Malariol., 1:* 211-220.

Puri, I.M. (1948) The distribution of anopheline mosquitoes in India, Pakistan, Ceylon and Burma, Additional records. 1936-1947. *Ind. J. Malariol.*, 2: 67-107.

Puri, I.M. (1960) Synoptic table for the identification of full grown larvae of the Indian anopheline mosquitoes, *Health Bull. No. 16*, Govt. of India, Calcutta.

Ramamurthy, V.V. and Ghai, S. (1993) Biodiversity and Conservation biology in Agroecosystems: An entomologist's Biosystematic point of view, *Hexapoda*, *5*(2): 127-131 pp.

Rajamany, R. (1993) Remarks on Insect biodiversity, *Hexapoda*, *5*(2), 1-4.

Rao, U.R. (1996) Science and Technology for achieving Food, Economics and Health Security, Presidential Address, I.S.C.A. Patiala, 1-48 pp.

Rao, R. T. (1975) Arboviruses and their vectors in India, *Indian J. Med. Res.*, *63*: 1219-1234.

Rao, R. T. (1984) *The Anophelines of India*. Malaria Research Centre, Delhi, I.C.M.R.; 518 pp.

Reid, J.A. and Knight, K.L. (1961) Classification within the Subgenera *Anopheles* (Diptera : Culicidae), *Ann. Trop. Med. Parasit.* *55*: 474-488.

Reid, J.A. and Weitz, B. (1961) Anopheline mosquitoes as vectors of animal in Malaya, *Ann. Trop. Med. Parasit.*, *55*: 180-186.

Reid, J.A. (1966) Anopheline mosquitoes of Malaya and Borneo Studies from the Institute for Medical Research, Kaula Lampur No. 31, 22 : 529 pp.

Reid, W.V. and K.R. Miller (1989) Keeping options alive: The scientific basis for conservation biodiversity, World Resources Institute, Washington, D.C.

Ross, Edward S., and Roberts, H. Redclyffe (1943) Mosquito Atlas: Part I. The Nearctic *Anopheles*, Important Malaria Vectors of the Americas and *Aedes aegypti*, *Culex quinquefasciatus*. *N*, 44 pp. Part II. Eighteen Old World Anophelines Important to Malaria. *N*, 44 pp. Philadelphia *Amer. Entomological Soc.* Washington.

Ross, E.S. and Roberts, H.R. (1943) The old world showing relation of species. List of Anopheles mosquito Atlas, Part I, II and III, *Amer. Entomological Soc.*, Washington.

Roy, D.N. and Brown, A.W.A. (1970) Entomology 3rd Edn. Bangalore Press, Bangalore, 855 pp.

Russell, Paul F., Rozeboom, L. E., and Stone, Alan (1943) Keys to the Anopheline mosquitoes of the world with notes on their identification, distribution, biology and relation to malaria, Philadelphila *Amer. Entomol. Soc.*, Washington, 152 pp.

Sabin, A. B. (1952) "Research on dengue during World War II," *Am. J. Trop. Med. And Hyg.*, *1*: 30-50.

Sathe, T.V. and Girhe, B.E. (2001) Biodiversity of mosquitoes (Order : Diptera) in Kolhapur district, Maharashtra, *Proc. Nat. Sym. Trends. Environ. and Evol. Biol.*, EB. AB, 27 pp.

Smith, G.V. (1973) Insects and other arthropods of medical importance, 37-107 pp.

Soper, Fred L. (1958) "The 1957 status of yellow fever in the Americas," *Mosq. News, 18*: 203-16.

Stone, A., K. L. Knight and H. Starcke (1959) A synoptic catalogue of the mosquitoes of the World (Diptera : Culicidae), The Thomas Say Foundation. *Entomol. Soc. Am. 6* : 358.

Strickland, C. and K.L. Choudhury (1927) An Anopheline survey of the Bengal districts. *Ind. J. Med. Res., 152*: 377-426.

Subbarao, S.K., Vasantha, K. and Sharma, V.L. (1988) Cytotaxonomy of certain malaria vectors in India, Clarendos Press, Oxford. 25-37 pp.

Tanaka, Kazuo, Kiyoyuki Muzusawa and E.S. Saugstad (1979) A revision of the adult and larval mosquitoes of Japan and Korea (Diptera : Culicidae). *Contrib. Amer. Entomol. Inst., 16* : 1-17.

Thakur, C.P. (2000) A Gloomy scenario; Family Planning Association of India, 22 : 14 pp.

Vitousek, R.P.R., Ehrlich, A.H. and P.A. Matson (1986) Human appropriation of products of Photosynthesis. *Bioscience, 36* : 368- 73.

Ward, R.A. (1984) Second supplement to "A catalog of the mosquitoes of the World" (Diptera : Culicidae). *Mosq. Syst., 16*: 27-232.

Ward, R.A. (1992) Third supplement to "A catalog of the mosquitoes of the World" (Diptera : Culicidae), *Mosq. Syst. 24*: 177-230.

Wattal, B.L. and Karla, N. (1961) Regionwise pictorial keys to the female Indian *Anopheles, Bull. Nat. Soc. Ind. Mal. Mosq. Dis., 9*: 85-138.

William, B. H. and Maurice, T. J. (1961) *Medical Entomology*, 5th Ed. The MacMillan Company, New York, 362-395 pp.

William, L.L., Jr. (1937) "Mosquitoes and Malaria", *J. Econ. Entomol., 30*: 20-26.

Index

A

B

C

D

E

F